TIMBER

WHIP-SAWING LUMBER

TIMBER

Toil and Trouble in the
BIG WOODS

by

Ralph W. Andrews

WHIPSAWING BOARDS in Alaska's early
days. (Uni. of of Wash. photo).

Schiffer Publishing Ltd

1469 Morstein Road, West Chester, Pennsylvania

Originally published 1968 by Superior Publishing Company
Copyright 1984 © by Ralph W. Andrews.

Printed in the United States of America.
ISBN: 0-88740-036-1
Published by Schiffer Publishing Limited, Box E, Exton, Pennsylvania
19341

This book may be purchased from the publisher.
Please include $1.50 postage.
Try your bookstore first.

DEDICATED

to Megan, Corrie, Andrew, Gregory and Barrie, young pioneers of the New West, who seem to push Yesterday back even farther.

Foreword to the 3rd printing

This 3rd printing of *Timber* has been improved by the author and new publisher to be the best presentation of the material so far. The illustrations have been reproduced for this edition from the Darius Kinsey photographs and the printing is on high quality paper to make the images stand out clearly.

My first book, *This was Logging,* presents Darius Kinsey's magnificent photographs selected to emphasize the commercial aspect of the Northwestern woods during the heavy lumber production years.

In this book, *Timber,* I have tried to impress upon people today, what it must have been like to set up a new life among trees of unbelievable size, in forests of unbelievable depth, and with axe and saw make the first attempts to wrest a living from them. The great masses of trees were at once terrifying and depressing yet they held hope of the abundant life. They were "toil and trouble" yet they exerted a profound, perhaps even a spiritual force, as they did upon the north coast Indians, and provided the white settlers with a common bond of interest, protection and industry.

Aiding in this pursuit were the Seattle Public Library; Robert Monroe, head of Special Collections Department of University of Washington; Oregon Historical Society in Portland; Vivian Paladin, editor *Montana Magazine;* University of Idaho; James M. Bogyo, Public Relations Department, Crown-Zellerbach Canada Limited; City Archives, Vancouver, B.C.; Provincial Archives, Victoria, B.C.; United States Forest Service offices in Missoula, Montana, and St. Maries, Idaho. I publicly thank all the people in these organizations for their willing cooperation.

RALPH W. ANDREWS

CONTENTS

And the
FOREST
covered the land

"One minute he was there. The next when I looked around he was gone and then he poked his head around the big tree six feet away where he was hiding and I couldn't see him. Trees! Such monsters, all crammed together as thick as corn stalks. God put them there, he must know what for." So wrote pioneer wife Amantha Sill as she started life in the Oregon coastal forest in 1861 — appalled, numb with unbelief, frightened. "It is a great cave for animals to live in, and Indians to come up on you so sudden. This is to be our home. Every day and night I pray to be taken back to Indiana."

South and north of the Columbia River, farther north to the Queen Charlotte Islands, Hecate Strait and beyond, the forest lay like a dark phalanx guarding the reaches from mountains to ocean shore against all intruders. Here in this seemingly limitless depth broken only by water courses and a few lush meadows, stretched the most magnificent body of timber the world has ever seen in this geological age.

Here was the rearing of centuries, without showing of youth or change, a silent, brooding and mysterious realm. Interminable, all but impenetrable, the wilderness allowed sound only to a few bird songs and the drumming of their wings, the staccato rapping of woodpeckers on hollow, rotting stumps, and to voices of tree branches whispering down with stray bars of sunshine from the crowns of columnar trunks reaching skyward.

"The dreary continuity of shade," said Charles Nordoff in *Harper's New Monthly Magazine* of February, 1874, as he referred to the forest encircling Astoria. "It had, I confess, a gloomy, depressing influence. The fresh, lovely green of the evergreen foliage, the wonderful arrowy straightness of the trees, their picturesque attitude where they cover the headlands and reach down to the very water's edge, all did not make up for their weight upon my sensibilities."

What kind of man could move among the masses of firs, cedars, hemlocks, the spruces, all rooted in antiquity, all so far removed from human activity, and not feel a stirring in his spirit? Sweetened by the sun, washed by driving wind and stifling mist, the sturdy boles inched upward in profound achievement while wars bloodied the earth and kingdoms tumbled to ignominy. Beyond the consciousness of majesty and veneration the forest transmitted a sense of ancestral memory. James Lane Allen wrote that within us is a vague awareness of this, of our relations to forest memories, one

"I REMEMBER, I remember the fir trees dark and high; I used to think their slender spires were close against the sky" . . . . . Thomas Hood. (Ore. Hist. Soc. photo).

11

"... so powerful that since the dawn of history millions have perished as forest creatures only; so powerful that there are still remnant races on the globe which have never yet snapped the primitive tether and will became extinct as mere forest creatures to the last; so powerful that those lightest races which have been longest out in the open — as our own Aryan race—have never ceased to be reached by the influence of the woods behind them; by the shadows of those tall morning trees falling across the mortal clearings toward the sunset."

Perhaps the haughty Spaniards who sent their small boats ashore were as awed by the forest as any raw settlers when they allowed themselves to be attacked by Indians to whom the trees were Spirit and Life. And perhaps the English explorers, with less of a desire to conquer than to observe and chronicle, were struck by the deep mystery of the forest and repelled by its forbidding front and oppressive depths.

Yet one point is certain. To none of the voyagers who first saw and to none of the intruders who braved the forests and lived in them, did they weave patterns of romance. They were either something the people could not understand or enemies which must first be routed, for this was a bone-brittle, earnest world in which there was no space for poets.

There was no Joyce Kilmer to glorify these trees and think them lovelier than poems, no Bryant to write, "The groves were God's first temples". Rather it was the lament of weary, hungry travelers like Samuel Bowles who in 1865 clawed his way through the western forest and wrote in *Across The Continent,*

"And off we bounded into the woods at the rate of three to four miles an hour. Most unpoetical rounding of our three thousand miles of staging in these ten weeks of travel, was this ride through Washington. The road was rough beyond; during the winter rains it is just impassable, and is abandoned; for miles it is over trees and sticks laid down roughly in swamps; and for the rest ungraded and simply a path cut

HEMLOCK. DOU

S FIR.   SPRUCE.   RED CEDAR.

"THERE WAS BEAUTY, yes . . . . but who would know until men found the forest and judged it so." (Uni. of Wash. photo).

13

through the dense forest — the height and depth are fully equal to the length of it. . . .

"These are the finest forests we have yet met, — the trees larger and taller and standing thicker; so thick and tall that the ground they occupy could not hold cut and corded as wood; and the undergrowth of shrub and flower and vine and fern, almost tropical in its luxuriance and impenetrable for its closeness. Washington Territory must have more timber and ferns and blackberries and snakes to the square mile than any other State or Territory of the Union. We occasionally struck a narrow prairie or thread-like valley; perhaps once in ten miles a clearing of an acre or two, rugged and rough in its half-redemption from primitive forest; but for the most part it was a continuous ride through the forests, so unpeopled and untouched, that the very spirit of Solitude reigned supreme, and made us feel its presence as never upon Ocean or Plain. The ferns are delicious, little and big,—more of them and larger than you can see in New England,—and spread their beautiful shapes on every hand. But the settlers apply to them other adjectives beginning with d, for they vindicate their right to the soil, in plain as well as forest, with most tenacious obstinacy, and to root them out is a long and difficult job for the farmer."

And faced with long and difficult jobs, the transplanted farmers, those early settlers, looked at the big trees as stubborn stalwarts standing firmly in their way of life. There must be food to eat and that meant growing grain and vegetables and they said cut the trees — use them or be used by them. The animals and Indians will go and our souls will be lighter.

When their axes bit into the bark there was no place for conscience. Few in that vanguard sworn to hack away the offending trees realized they were bringing noble lives to an end. And those who did regret rationalized on the necessity of it and left the weeping to the sentimentalists, the statistics to the historians and naturalists.

Later they would read of fir and spruce trees like the ones they had taken down being seedlings in the sixth and seventh centuries when the Jutes, Angles and Saxons were overrunning Britain. The forest service men would tell them trees six and seven feet in diameter were very common in the northwest forests and that there were thousands of areas supporting fir and cedar trunks reaching eight feet and larger, three hundred feet and more high.

With swelling pride the naturalists saw only beauty in the forests, the trees as symbols of peace, content and protection, and at first considered them inexhaustible. Essayists and philosophers echoed their praises. Washington Irving declared in *Bracebridge Hall,* "There is an affinity between all nature, animate and inanimate." And Emerson said, "At the gates of the forest the surprised man of the world is forced to leave his city estimates of great and small, wise and foolish. In the woods we return to reason and faith. There I feel that nothing can befall me in life, no disgrace, no calamity (leaving me my eyes) that nature cannot repair."

Such friends of the forest found the Douglas fir the stout leader of its fellows, lofty, kinglike in its age-old majesty, its legions gathered closely around it. The red cedar was its tender consort, more delicate, showing more elegance, finding deeper roots in the bottom lands and river areas to give it great endurance. To the spruce was given less aspiration to height, girth but a wealth of grace, a tree content to

". . . lift its mighty pillar straight and direct toward heaven, bearing up its honors from the impurities of earth, and supporting them aloft in free air and glorious sunshine, an emblem of what a true nobleman should be — a refuge for the weak, a shelter for the oppressed, a defense for the defenceless, warding off from them the peltings of the storm or the scorching rays of arbitrary power."

LIFE AND DEATH — 30 year old hemlock seeded itself on long-dead fallen cedar and one life went on as other fell deeper into limbo. (Ore. Hist. Soc. photo).

SYMBOLS OF PEACE, contentment and protection; useful servant and faithful friend of man. (Ore. Hist. Soc. photo).

TIMBER — Vancouver Island. (B.C. Prov. Arch. photo).

SUNLIGHT seeps through foliage of group
of hemlock trees in Oregon. (Ore. Hist. Soc.
photo).

STURDY HEMLOCKS in Skagit County,
Wash. (Uni. of Wash. photo).

GRIM WERE THE FACES and sore were the muscles of the settlers who took out the trees to open up the wilderness. (B.C. Prov. Arch. photo).

First Axes
in the TIMBER

The rain fell steadily, day and night, not borne on the wind but sheeting down gently without ceasing out of a gray sky "like a dirty galvanized bucket turned upside down" as Wilmer Hutchins wrote in a letter to his father. "Everything was soaking wet, water running down the trail I made into the timber like a little creek and my shack leaked in a dozen places. I'm learning fast how the trees grow so big here in Oregon."

The newcomer from the farming east or prairie states might well have fingered the edge of a double-bitted axe with the feelings of an inmate of some labor camp condemned to tearing down a mountain. Or he could have suffered black discouragement as he watched others, longer settled on the fringes of West Coast timber, mount spring boards ten or twelve feet up the trunk of a mammoth cedar, well above the swelled butt, and drag a long, heavy cross cut saw across it with dogged patience until it crashed down and was split up for shingles.

Then perhaps to his pleased surprise he learned that an able man, by working from dawn to dark, might cut and split a cord of those bolts and swamp out a path so the chunks could be dragged "out to daylight" in a stone boat. And get a silver dollar for them. If he had claim to the timber he might sell the bolts or split shingles for twice that or more if he had a horse or boat to haul it to some settlement. Life he knew was "not without dust and heat".

Living meagerly, dressed poorly in whatever clothes he brought west with him, the first year pioneer settled grimly to what he had to do. He took it by the day and every month he could see his hole in the woods growing a little wider as he sunk the family roots deeper in the rotted-wood soil. If he faltered at times a voice or two at his sleazy cabin, or in his own conscience, tugged him back up the mucky trail to where he had left "that confounded axe" and splitting tools.

It was something like that with Wilmer Hutchins. His first winter lengthened out long weeks at a time as discouragement beset him. After the back-breaking work of cutting down trees, dragging short logs to the beach where he could borrow the use of a horse belonging to his only neighbor William Littlejohn, he was only a knife-edge away from giving up and joining his brother on a land claim farm in the Willamette Valley. Starved out of the California mines the two Pennsylvanians came to the Columbia River country on a sloop, Wilmer leaving it at Astoria, brother Evan going on up to the Willamette.

1852 it was now, the first winter spent in taking down the noble firs and cedars. When the rains eased up Hutchins and Littlejohn went partners in a whipsawing project, building the frame and splitting the logs into workable sizes. They had sent to San Francisco for the heavy saw and by mid-summer had a stout pile of fir boards which they carted to Astoria behind the horse on a pair of wheels made from 2-inch slabs sawed from a log 3 feet in diameter. The boards brought enough for salt pork, flour and dried apples to help keep the two alive another three months and some big nails needed for the building of a storage shed for lumber. The storekeeper told them to send an Indian when they had another lot of boards ready and he would bring a boat around for them.

"Indians. Well they were friendly enough," he told his father in a letter, "but no good to work and all they want is to beg for food and old clothes. They have a little village along the beach aways, we can smell it when the wind gets around in the north. Awful dirty they are. Some women came and gave us some clams once, a young one and two old squaws and I would sooner have a dog around than them. The girl was awful curious about William and it wouldn't have taken much to get her to bed but he decided like me, it was better to shoo them out than have some canoes full of Chinooks raid our camp. There must be an acre of rotting clam shells making that smell."

Hutchins was 26 that first fall and full of doubts about his judgment in coming out west. Yet at Christmas time when he wrote his father again he thought he would stay. He got on fine with William Littlejohn he said and maybe they would build a water-powered sawmill on the creek the next spring when they could get the storekeeper to help them with some money for it and give them credit. "My experience as a carpenter with you is a big help and I know most about putting up that sawmill.

"Up that creek there are some cedar trees bigger than twenty like you ever saw, maybe 250 feet tall and it is wet in there all the time, the sun never get down in only in little slits.

A quarter of a mile in there is a watering place for animals and when we need meat we take a day and go to that hole early in the morning and shoot our pick of six or seven deer. There are a lot of broken down trees laying down in and across the pond and William saw a mountain cat drinking under one just at dusk. He got up and away before William could get a shot at him.

"People say the forest is too lonesome a place to live in, they have to be out in the open, but not me. It is the greatest you could ever imagine only awful quiet. No birds and no squirrels that I ever saw but lots of bear and deer. The trees are so big some of them must be 300 years old and big cedar logs laying on the ground covered with moss and fungus, the wood is still good.

"There have been some big winds here and you can tell some of those big trees blew over and some went by fire too as there are a lot of black spikes and charred trees in one part south of where we are. A man up at Astoria told me the Indians said there was a big fire down that way that was started by lightning and destroyed a lot of trees and burned for a long time. When I said we were the first white men in this part, he said no, there was a cabin right near us and we should have found it. So William and I set out and scoured all around and sure as anything we found it, pretty tumbled down, but who lived there we don't know, not the man in Astoria either. But maybe a trapper. It was five years old William and I conjured."

And no doubt in a few years other settlers found the Hutchins-Littlejohn buildings and wondered who had lived there, for the next winter they were gone and the record ends. They might have been killed or taken sick or fell victims to the hard work or perhaps the pull to the brother's farm proved too strong for Wilmer and he took his partner with him to the Willamette.

Other axes were ringing up and down the Northwest Coast. In 1856 Capt. Alexander Sampson of Duxbury, Mass. sailed his ship up the Juan de Fuca Strait, trading with the few settlers and Clallam Indians along the southern shore. He reported finding a widow living with her son, about 16 years old, who had suddenly become a man when his father died. Their log cabin was on a claim back in the timber where they had intended to establish a homestead. The mother was fast becoming infantile and the boy scratching out a living helping other settlers cut down trees for market. In one of her irrational periods, the mother sat at the cabin door with a rifle and in the dusk mistook her son for a prowling Indian and shot him.

Many of the vessels that brought settlers to Oregon, Washington and British Columbia stayed to found settlements, using the ships to carry meager cargoes of lumber to San Francisco and bring more settlers north. The schooner *Albion*, flying the British flag, hove to in Discovery Bay as the skipper considered the tall, straight firs and spruces as ideal material for ship's masts. When he sailed away he had contracted for a deckload of peeled trees and four months later he brought the schooner back to load them — seventeen spars, 70 to 100 feet long, over 2 feet in diameter tapering to 18 inches.

The historical record gives the ship's carpenter, William Bolton, as the man in charge of the cutting crew, mostly Clallam Indians who worked however well for clothing, tobacco, ornaments, fish hooks and knives from the *Albion's* trading stores. Bolton was properly impressed

"MY HEART sank to some awful depths
when I realized all the hard work it was
going to take to get us firewood and a place
to farm." (Uni. of Wash. photo).

"THESE WERE SILENT WOODS, mostly. Living creatures seemed awed by the majesty of the cathedral arches overhead, the columnar trunks reaching skyward." (Ore. Hist. Soc. photo).

with the resilient toughness of the fir wood, heavy resinous sap and the sheer size of the trees — so big, he said, he had to look hard to find ones small enough for the intended purpose. But find them he did, had the limbs chopped off, the trunks squared and gangs of Indians pressed into dragging the bulks to the beach. Even so, it took days to haul one log, the manila hawsers parting many times.

Piling and ship spars were being cut around Puget Sound and the Gulf of Georgia during the '50s and '60s. Early any morning in those years smoke could have been seen rising from crude chimneys of scattered cabin homes on Alki Point in Puget Sound where the vanguard of Seattle's founders had settled. The children were up, women cooking fish, clams or game from the forest with the rudest equipment, later making clothes and mending for the families. The men went out early to cut down trees for masts and spars, barking them but leaving the logs where they fell to be dragged to the water with the help of the ships' crews. They blazed trails, grubbed out the clearings, slashing and burning brush, or hunted for deer and grouse.

And always the rain persisted, renewing the forest as fast as the few axes cut it down. The Coos Bay country of Oregon was a veritable water wonderland with rivers, inlets and sloughs fingered into the dense timber. Settlers quickly learned that the Port Orford cedar growing profusely in the area far exceeded any other wood for boat building. It had all the enduring qualities of western red cedar, was in addition, impervious to acids, termites and salt water teredos.

The first Coos Bay inhabitants used this cedar for their own small boats to navigate the intricate channels in maintaining contact with their neighbors and for split shakes used to roof and face the cabins. A general store was set up here in the late '50s by Abner Watt, a quartermaster from a British ship. To take up his spare time he made a partnership arrangement with young Carse Elliott who had worked his way through the forest from the Rogue River area, following the winding course of the Coquille River. Elliott and Watt were to peel the chittam trees, dry the bark and hold it at Watt's store until a San Francisco-bound ship would pick it up. The bark was valuable, used in the preparation of the laxative cascara.

Watt was a gruff, grisled seaman, barrel-chested and balding, stubborn and irascible. He slept in a shed tacked on to his store, the earthen floor usually muddy from the leaky roof. It was said he offered to pay his partner Elliott to re-roof it with good cedar shakes but the work was never done. Watt slept on a bunk made of knotty tree branches and cooked his simple meals outdoors or munched them cold in the store.

Stacks of chittam bark covered the barrels of coal oil and sacks of potatoes and the two men were starting to build a storage shed when Elliott disappeared. He had many friends among the settlers and they seemed to find Watt's answers as to his whereabouts vague and misleading. Then one Anson Spriggs went fishing in the bay and saw what looked like a man's hand raised a few inches above the low tidewater under Watt's dock. He peered into the water and decided the body was Carse Elliott's, badly decomposed. Securing it to a piling he waited for a lower tide the next morning and took two men with him to raise the body. It was that of Elliott and they pulled it up into Sprigg's boat. One of the three was armed and they confronted Watt in his store, demanding an explanation. The storekeeper had none, only a grumbling disavowal of any knowledge and ordered the men out. Instead of going the armed man started toward Watt who knocked the gun aside and charged like a bull. Spriggs snatched up a bacon knife and jumped into the melee. Watt knocked him down, wrenched the knife away and was about to run it into Spriggs' chest when shot dead by the man with the deer rifle.

The affray was duly reported to the captain of the next ship arriving but no action of law was ever forthcoming. The store and bark picking activity was operated as a community enterprise as was a small sawmill built by Spriggs who later acquired a small schooner to ship the sawn boards and bark to San Francisco.

On the passenger lists of ships sailing north from the Golden Gate were many disgruntled '49ers leaving the gold fields for greener ones, and in the welter of displaced persons was a generous salting of adventurers, outright criminals, flimflam artists, rascals and rebels without causes — and some New Thoughters with them. One such group was the Servants of the Lord, a rag-and-bones gathering of religious cotton-heads herding together to evade legal and social restrictions, hanging to the frocked coat of a man

"TIMBER, TIMBER 'til you can't sleep!"
The great forests made a profound impact
on the pioneers who had never seen or
heard of such heavy growth of timber. (Left,
Ore. Hist. Soc. photo — right, B.C. For.
Serv. photo).

using the name of Henry Gilson, a psalm-singing circuit rider from the Ozarks. When he told the group they were sailing to Portland on the British bark *Merridew,* that he had paid passage for twenty-two of them, they could only salaam and praise the Lord from whom all blessings flow.

All except Chauncey B. Veit. He may have mumbled grateful words and looked with spaniel eyes upon leader Gilson but he had the heart of a jackal and lived in daily hope that some fat cumshaw would come his way. He had little respect for primitive law in any land or fraction thereof and once landed on the Columbia shore he set about to prove it.

Gilson and company were moved up the river by Chinook Indians on a trading mission and established a colony up what was later named the Wind River. They took vows to make the enterprise prosper., the men by cutting into the heavy growth of pine and fir, floating it down to the Columbia. Getting it across that turbulent current was a matter which the Lord, who moved in wondrous ways, would take care of, like sending the seagulls to save the Mormons.

A heterogeneous group like the Servants were never safe from trouble, either from tempers or hunger and while Henry Gilson had genius along certain lines, he lacked it sadly in another. He talked too much. And right there to pick up dropped confidences was the scheming Mr. Veit. In fact he had wormed his way into Gilson's inner thought passages by nearly working himself to death in the woods to show the sweat of honest endeavor and by spelling the leader in sunset vespers. Then on one starry night he maneuvered the bearded prophet into telling him where he got the money to finance this freedom-for-all venture and where he kept it. Henry Gilson felt that it was only right to tell and that was fatal. Chauncey Veit found the cache of some four thousand dollars in a spare boot, one of a pair always thrown carelessly in a dunnage bag along with pans, old clothes and some moldy Bibles.

When the group woke up the next morning it was without Veit. Crushed at losing "his right arm", Gilson was horrified to discover the money gone and damned Chauncey B. Veit into all the caverns and crevices of hell, seeing the Servants of the Lord and himself sinking into the mire with every curse. The colony was dissolved and if Veit ever crossed the prophet's path history neglected to make a record of it, perhaps because the crossing was so short and the remains unidentified.

Flagpole Epic

All that scary talk! Why these northern Indians in Washington Territory were so friendly they were going to show him how to put up a temporary camp in the forest against the wet winter weather and even help him and his boys build it. The plains Indians were warlike and troublesome but these fellows pointed out a big fir tree where he could put up the camp and by Jupiter, he'd call it the "Pioneer Tree"!

So did the first family start life on the hill called Claquato, a high place, by the Indians on the south side of the Chehalis River. They hacked out a crude road to a shallow place in the river where they could ford it and haul in the household goods brought up from Portland.

This was the Lewis Hawkins Davis family from Vermont and Indiana. In Portland Davis heard much talk of good farm lands available farther north so he and family set out by boat down the Willamette to the Columbia River. At its junction with another tributary, the Cowlitz, the party went upstream to Cowlitz Landing. The surrounding country looked good to pioneer Davis but he selected an area centered by a rise of land and took out a donation land claim.

By 1855 the Indians had turned bitter, realizing that unless the increasing white tide was stemmed, they were doomed. Frequent skirmishes with settlers all over western Washington caused the government to authorize the building of blockhouses as protective measures. At Claquato Davis got the local contract, providing manpower and logs. The structure was built on the brow of the hill just west of the settlement center, the always hungry workers fed by Mrs. Davis from a kitchen set up on the site. When the building was finished all families moved in but after many quarrels natural to strangers in cramped quarters, they returned to their homes.

The first settlers built cabins of logs cut from the smaller trees. The hill was not solidly forested and opening it up increased pasture area. In 1857 Davis built a whipsaw frame at the bottom of the hill, a pool in the creek enlarged to hold logs skidded in by ox team. A dedicated Methodist, Davis used the first lumber to build a church.

While the Civil War was raging, although it hardly whispered into this northwest corner, the Claquato women were resolved to be patriotic and make a flag that would be a credit to the newly built courthouse. John H. Browning, Davis' son-in-law and owner of the general store, was in San Francisco buying supplies when he received a letter from his wife telling him to bring home materials for a flag 18 by 36 feet. Although convinced the size was in error he returned in mid-April with $90 worth of cloth and thread. In his home the women hand-stitched thousands of feet of seams to complete the giant emblem. And a crew of men scoured the forest to find a big tree of proper proportions for a flagpole. Trimmed, peeled and dragged home, the pole was placed in a 20-foot hole and then stood 120 feet.

The war was demanding immense quantities of medical materials called "sanitary supplies" and an inspired suggestion was approved, to call the courthouse dedication the "Sanitation Ball", all proceeds to go to the war effort. The big day came on the Fourth of July, 1862. At 10 a.m. minister John Harwood opened the ceremony with prayer. Thirteen men, one for each of the original colonies, fired four volleys and the flag was slowly raised to the top of the pole while the people sang a hymn. After a picnic there was a dance by the river to the music of an orchestra, ending in a blaze of "Old Glory". Tickets sold at $5 each and after all expenses there was $250 for the sanitation fund. The flag was flown every day for the remainder of the war but on October 22, 1864, it was lowered to halfmast for Lewis H. Davis and again in 1865 for Abraham Lincoln.

The Indian scares of the middle '50s that brought tragedy and death to other Washington communities never directly affected Claquato and the building of the fort was entirely futile. Local Indians became friendly and cooperative as in the founding days, small groups camping nearby for many years. Probably the best remembered native was "Queen Susan", wife of a powerful chief in the area. She was still young and attractive when he died and could have married a brave of high rank, but she fell in love with a slave. She married him and out-

"WE FOUND A CRUDE CABIN, already sagging to its earth floor, where two families were struggling to scrape out a vestige of secure existence..." (B.C. Prov. Arch. photo).

raged members of the tribe rejected her, but as royalty, permitted her to remain, with the same lowly status as her mate. When he died the middle-aged and fat widow lived among the whites in Claquato. Dressed in brilliant, cast--off clothing she went from house to house doing any kind of cleaning or laundering chores for a living. Her cheerful nature was a bright light in the community.

Susan died in 1868 and the citizens gave her a funeral as imposing as one the tribe would have given her as queen. Most of the local people attended and even the passengers of two stages which stopped for a change of horses witnessed the rites. All Indians living within several miles also came, having forgotten their queen's descent from grace. Susan's body was taken to the foot of the hill and placed in the Indian burial ground.

EARLY SETTLER on Shawnigan Lake, B.C., W. E. Losee, was able to finance a sawmill. (Gerry Wellborn photo). Below, Gilley Bros. in New Westminster, B. C. had pioneer supply and trucking business. (B.C. Prov. Arch. photo).

"TALL FIRS, straight as an arrow, dense as a southern cane brake" — Vancouver Island. (B.C. Govt. Trav. Bureau photo).

YOUNG FAMILY at Ft. St. John, B.C.
who found "a new country practically as
the hands of its Maker had left it. . . . ."
(B.C. Prov. Arch. photo).

(Opposite) "FIRS SO THICK and tall the
ground they occupy could not hold them
cut and corded as wood" . . . . (B.C. Prov.
Arch. photo).

"THESE PEOPLE we found in sturdy
quarters were not in any sense adventurers
but they were men and women of stamina."
(B.C. Prov. Arch. photo).

(Opposite) FARMER-LOGGERS in Washington's big timber. (Uni. of Wash. photo).

Oysters in Timberland

In one small Washington Territory area on the Pacific Ocean it was a question which came first and which was most valuable to the early settlers there — the timber or the oysters. Both were there on Shoalwater Bay when the first white man arrived, the south arm a vast field of oyster beds.

This was the small, delicately flavored *ostrea lurida* which thrived here under ideal conditions for reproduction and growth. It required an almost perfectly formulated mixture of fresh and salt water, and the saline content must vary at specific intervals. Shoalwater Bay offered an inflow of fresh water from the Naselle and other rivers, while the tide was at full or receding which

alternated with ocean flood tides. Native Chinook tribes had traditionally used oysters as a basic food but had made no impression on the supply.

The first white man to see commercial possibilities here was Virginian Charles Russell who discovered that during the time the non-ambulatory oyster was exposed by outgoing tides it closed its valves tightly which excluded desiccating air, and that this short period could be extended, which meant they would be edible for many days after being picked.

The only entrance to Shoalwater Bay was at the north end, some 30 miles north of the Columbia River from which Russell decided to ship his oysters to San Francisco. In prehistoric

SUMMER AND WINTER log hauling in the north country. Above, western soft pine near Nelson, B.C. and opposite, short logs on 4-horse sled at Eagle Lake near Giscombe, B.C. (B.C. For. Serv. photos).

times the Columbia had emptied at least part of its waters into the now separated northern estuary, the old channel now low and marshy, a route used by Indians as a portage from Columbia to Bay. In the summer of 1851, with a partner, Russell took a canoe over the passage and at low tide walked over the flats, easily collecting a sack or two of oysters which he took back to Astoria. The lot reached San Francisco in good condition and was sold at a good price.

Heartened by visions of success Russell put into action ambitious plans to improve the southern access route to the source of supply. Meanwhile he learned a Capt. Fielstad had run a schooner directly into Shoalwater Bay, followed closely by the *Sea Serpent* and *Robert Bruce*. They all sailed south with oysters, one delayed by storms while cargo spoiled and one burned to the water's edge by a mutinous cook. But the oyster rush was on and with the drive of R. H. Espy of Wisconsin who arrived in 1854, and I. A. Clark, the town of Oysterville prospered until the late 1890s when the larger South Bend was awarded the courthouse and commandeered all records from the settlement oysters built.

So the timber had the last laugh, yielding up only a few trees for building purposes until loggers took it by storm, long after the oysters and the settlement were part of early history, the bay renamed Willapa.

"IN THOSE DAYS", wrote a doughty British Columbia settler, "we hired anybody we could get to clean out the woods and sell the logs to the little water powered sawmill." (B.C. Prov. Arch. photo). Below, 75 years later, logging was still done the hard way. (B.C. For. Serv. photo).

38

SKIDDING LOGS to the homestead for the
winters' wood. (B.C. For. Serv. photo).

LOADING WESTERN LARCH on sled in British Columbia. (B.C. For. Serv. photo). Below, pioneer indicates log diameter is over 8 ft. (Ore. Hist. Soc. photo).

"BEAUTIFUL TREES? Yes," said Samuel Bowles, "but the settlers apply to them other adjectives beginning with a 'd' for they threaten their right to the soil . . . and to root them out is a long and difficult job . . ." (Uni. of Wash. photo).

"MEALS" reads the sign on this settler's cabin by the side of the brushed-out trail through the woods. (B.C. Prov. Arch. photo). Opposite left, masts for ships were ready to be cut, straight as arrows, no knots for 100 feet, 40 inches in diameter at 30 feet from base. (B.C. For. Serv. photo).

NEW OREGONIANS have sniped logs ready to haul them out of deep timber with horses. (Ore. Hist. Soc. photo).

EIGHT TOUGH TIMBERMEN stand in undercut of tremendous fir near St Helens, Ore. (Ore. Hist. Soc. photo).

PIONEER STEAMBOAT built of whipsawed planks on British Columbia lake. (B.C. For. Serv. photo). Below, long day's work starting for farmer set to remove fallen tree. (Ore. Hist. Soc. photo).

"ME AND A HORSE could clean out a patch of land," said one intrepid pioneer — but not this patch, not firs 6 and 7 feet in diameter. It took years to "let the daylight" into some heavy cover. (B.C. Prov. Arch. photo).

WHIPSAWING BOARDS at Finlay Junction, B.C. — daily output 150 linear feet. Beside pushing up on the saw, the man on the ground had to keep sawdust out of his eyes and mouth. (B.C. For. Serv. photo).

OREGON PINE woods east of Cascade
Mountains displayed park-like beauty to
early tourists. (Ore. Hist. Soc. photo).

FAMILY PARTY on giant fir felled by
George Crey (shown on ladder) in August,
1895, near Vancouver, B.C. Tree measured
417 feet in height, 25 feet in diameter at
butt, 9 feet at 207 feet from ground. Bark
near base was 16 inches thick. (B.C. Prov.
Arch. photo).

(Opposite top) PIONEER SAWING
PARTY on Dease River, B.C. (B.C. For.
Serv. photo). Bottom, cutting boat timber
on Lake Bennett, Yukon Territory. (Uni. of
Wash. photo).

DAYS OF '98 SAW MI

MONARCHS OF OREGON coastal woods
(Ore. Hist. Soc. photo).

BIG CEDAR on Oregon homestead, 15 feet in diameter at base. (Ore. Hist. Soc. photo).

BUCKING SAW built by enterprising mechanic for cutting fallen logs to be split for firewood. Below, bucking big fir in Oregon. Man on top apparently just came along for the ride. (Ore. Hist. Soc. photos).

LOGGERS
REST . . . 1870

A thousand feet back from the beach was the timber. Small firs and hemlocks these trees were and into the mass of them a road was cut. At the foot of it and spilling down over the rocky slope to the high tide line was a clutter of forty or fifty fir logs, ends sniped, bark half scraped off, the green wood scarred and mangy. And over the restless tidewater, the stony beach, the logs, the forest — gray-fronted in the dusk — over everything on the whole Northwest Coast rain fell in a steady, cold curtain.

The road was no ordinary one but crossed from side to side every eight feet by small logs half-buried in the spongy, root-webbed earth over which the logs on the beach had been dragged by the sheer power of six oxen. A few hundred feet up the grade was the camp clearing, smoke struggling up from the tin pipes of two large slab-sided buildings, oxen feeding under the sloping roof of a lean-to at one end of a small corral.

This was a logging camp in the first days of logging on the Oregon or Washington coasts, Columbia River, Puget Sound or north up Georgia Strait. The big trees were being taken down in earnest, driven down the rivers, boomed to sawmills. It was the first concerted effort to "let daylight into the swamp".

Life was plain and simple — work in the wet forests from dawn to dusk, sleep "like a stunned sheep" for eight hours, eat corned beef and cabbage, salt pork and potatoes, flapjacks and molasses three times a day.

Short stakers, long stakers, the loggers were single, tough and glad to get a dollar a day. They were Swedes, Finns, English, Irish and Welsh and "anybody that had ought to know better". They came to the camps with blanket packs, were hired as swampers, fallers, sawyers, hook-tenders, bull punchers or teamsters. The woods boss said, "See the bull cook" and the bull cook said "That's your bunk and if you want it softer get yourself some spruce boughs." The bunkhouse held twenty to forty of them built around the seven-foot walls, a boiler stove in the center of the room. The rain hammered against the wood walls, ran down through the pipe hole and sizzled on the hot stove. The air was thick with the acrid odor of wet clothes and socks hanging from lines strung from bunk to bunk and from a box of cedar chips near the stove, used as a spittoon and to catch a bad roof leak. And every time the bull cook noticed it he said, "I'm gonna clean that out some day."

He lit the oil lamps about 5 p.m. when the men came trooping in wet to the skin, stomping on the dirt floor. If there was a new hire they told him the first six months was the hardest and, "Where's the whiskey you brought? The new loggers fixed up their beds, spread their blankets and looked the company over, some with short beards, some long with hair falling down to their wool shirts. The bull cook filled the wash basins from a barrel outside and motions were made to wash off some of the sweat and dirt. A few men tried to read books or write letters by the dim light or ask the bull cook, "How many pigs you got out there now, Ambrose?"

The skid greaser came in with a new pair of socks from the company store, his howl about the price not heard above the booming babble. "Hell, Michigan was worse than this." . . . . "Come April maybe I'll go up to Canada in the pine." A lanky man with a six-inch mustache showed his collection of girlie postcards. On the wall of his bunk hung a calendar with a picture of a Japanese dancer and the dates marked in red as they went by. "Anybody got some shoe oil?" and a raspy voice shouted, "Mike — go tell that cook to sound that damn gong. I'm so hungry I could eat a bitch wolf!"

Inside the cookhouse dishes clattered as the flunkey jumped from table to table with the canned milk and doughnuts setting them on the flowered oil cloth, working up an appetite with every smell of the meat frying, pork and beans simmering. The cook wiped flour off his mouth and turned from the range to bark at him. Steam curled up from the soup pot. Two cats rubbed the cook's legs, mewing for supper. In came the bull cook to turn up the lamps, throw his coat under the kitchen table and back up to the stove to dry off. Then the cook walked majestically into the dining room to survey the scene. The flunkey ran to the door with a small bar of iron ready to sound the supper gong, to clang the "gut-hammer" on signal. "Everything ready?" asked the cook. "All dandy fine," said the flunkey. "Let her go," said the cook.

The door swung open and the men filed in with orderly patience as the cook eyed them with his arms folded under a steely stare. "That's your seat," the flunkey told a new man, "and see you don't sit nowhere else." Cups and plates were turned rightside up and hands reached for pitchers of milk and tea and coffee as platters of meat and beans and potatoes were set down. Then it was snatch and pitch in. Stacks of bread went down like leaves falling and even through the raisin pie there was no talking, no sound but the clatter of knives and forks and rattle of rain on the shake roof, chewing sounds and a belch here and there. The cook said no talking and what the cook said was law. Chairs scraped back and men straggled out, picking their teeth.

Two hours and a half 'till nine. A low-ball poker game was started, a short-bearded Finn totting up his winnings for the winter — 35 cents. A knot of men were logging it hot and heavy near the stove, rolling cigarettes, taking "a rare

of snoose" or two. "Like I told that hayneck — I don't mind you ridin' on your end of the saw only just don't drag your feet." On one bunk a man was fitting a patch on his boot, another brewing some cough remedy made from bark and berries on the stove. The bunk of one young logger was decorated with odds and ends in pack rat fashion. On a shelf made from a cigar box was a faded photograph of a ship with a straw-haired lady standing by it looking about as big as the ship. A tobacco tin near the head of the bunk held his matches and hanging on a nail was a big nickle-plated watch neatly encased in a round snuff box with one side partly cut away. The top blanket was red-checked and the pillow a neat green. A squirrel tail was nailed to the wall near the match tin.

A voice sang out, "Look at old Eric making that boat thing inside that bottle. Beats all hell why. I'd rather be gettin' something out of it. Whatsa matter, Joe?" The man he looked at was sitting on the edge of his bunk, eyes watery, chin hanging low, the picture of dejection. He was Joe McCook who had worked at least a week in every logging camp south of Kelsey Bay, British Columbia, to the Klamath River in Oregon. He knew about all the woods bosses, cooks and bull cooks in the northwest and answered to nicknames like "Short Stake McCook", "Ten Day McCook", "Boomer Joe" and "Camp Inspector".

Joe was sick. Twice he went to the cook-house to get some lemon extract "to settle his stomach" and came back empty-handed to sit on his bunk and watch some pre-historic animal slither up out of the water pail on the stove and fade away. But this was not worrying Joe as much as those bells ringing down in the woods somewhere and how soon he can get away from this godawful camp.

And what about A. K. Brennan? A. K. Brennan started work in the Maine woods 40 years ago. He used to blow in the stakes at Bangor and when civilization got too close, packed his turkey and moved on to Wisconsin. He cut white pine there and drank his wages in gaudy saloons in Merrill and Eau Claire. When the push to the west started he came with it to fall the big firs and cedars and cash his pay in the skid road joints on Cordova, Yesler and Burnside Streets and told yarns about saloon brawls, about snowbound logging roads so steep

CLATSKANIE LOGGERS in camp shown on following page. Men are holding copies of *Police Gazette*. (Ore. Hist. Soc. photo).

that bridle chains failed to slow the sleds down, about white water streams whose beds were strewn with the cantdogs of drowned rivermen. "Never mind the men but save the cantdog", was the order when a shanty boy fell in. "Men was plenty," A. K. said, "but a good cantdog cost two dollars."

Nine o'clock and nobody talked after the bullcook turned the lamps out — nobody wanted a corked boot in his face. And only a few groans at the bullcook's "Roll out!" at five in the pitch black, cold, wet morning. Roll out they did and had their eyes open half an hour later at breakfast. At six they reported to the woods boss who set the crews for the day.

The cook, flunkey and bull cook did their chores to the rain-muted singing of cross cut saws, ring of chopping axes and raucous bellowings of a bullpuncher trying to move a stubborn ox team. Sunday was different. That was the time men washed their underwear and socks, tried to discourage the new crop of bed bugs growing active since the bullcook had covered the ground and had haircuts if somebody was handy with the scissors. Or it was "a day in the hay", to read a book, write a letter or lie in the bunk and listen to the rain or dream of the day when a man could turn in his tools, get his pay and catch the steamer for the nearest town and "blow her in."

EARLY LOGGING CAMP near Clatskanie, Ore., probably Peterson's, about 1900, with quarters for about 100 men. Note bull corral at left, log cookhouse behind tree at left. (Ore. Hist. Soc. photo).

Wood Choppers' Camp

Woodland Camp, on the site of what was formerly the Lowe and Killene claims, is some twenty miles from Everett and supplies the paper mill at Lowell, one of the best known Washington manufactories, with fuel and material for "pulp", sending to the mill daily from ten to fifteen carloads of cordwood.

As the study of industries as well as of humanity under novel conditions is always interesting, the different phases of woodcraft not always possible to a town-dwelling woman, the caboose of the Woodland train, attached to a long line of flatcars with huge wood racks, was entered with a feeling of satisfaction, notwithstanding its being stuffy with a prevailing odor of leather cushions, which according to tradition and the assertion of appearances, had done duty as coaches for traveling loggers in various stages of illness or intoxication for a long period.

The conductor, addressed by his associates as "Stone", was genial, the brakeman attentive to his two passengers — the girl bound for some little mountain station in silky crepon with a yellow novel, and the woman in search of information of wood camps.

"You'll be cold out there; come in by the stove", he said cordially, ushering them into the forward compartment, which like the one just vacated, was not modeled on the lines of a Pullman. As wear and tear and a certain kind of comfort are all that can be considered on a wood train, the coal bin was built up from the floor of the caboose, close beside the long-necked, club-footed stove and a great shovel in the bin made easy the brakeman's task of keeping it red-hot.

With the object of the journey in mind, one accepted the surrounding as picturesque accessories to the wood industry; the thick soled boots dangling frankly from a hook, the yellow oilskin coat on the opposite wall, the iron washstand in the corner with its necessary litter of soap and soiled towels, the curtains a woman would have

This account originally titled "The Man With The Axe" by Clara Iza Price, appeared in the Seattle Post-Intelligencer of March 17, 1907.

draped back from the windows, but which were tied in the middle with a knot of masculine determination that couldn't come undone to the end of time.

There was a great deal of shunting from the mill to the station before the conductor and brakeman finally climbed to their swinging chairs in the observatories, opposite each other at the top of the coach, and the engine went puffing over the drawbridge into the forest.

For the first few miles the track runs past little ranches whose few acres have been hewn from the forest. These tracts are still thickly set with the trunks of the trees that a year ago rose 300 feet to the sky and shut out the sunshine from the fertile soil.

Soon glimpses of ranch life cease and on either side of the track appear only trees — trees, the base of their trunks hidden with the jungle-like growth of the Washington forest which prevents the eye from penetrating a rod beyond the outer edge of the wall of spruces with blue green spines and suggestions of brown cones, the cedar's dark plumes, the hemlock and fir.

At Hartford junction the train slacked speed and a man with a piratical beard and tender voice, accompanied by a little girl with bare, purple hands, got on board. Here the tracks of the Monte Cristo and Sumas roads make deep trails among the green of the woods, and the legend is real that Woodland is two and one-half miles farther on.

Presently a man with milk cans comes out of the woods, another slouches through the door of a saloon and boys with dogs appear from somewhere and climb over the high steps of the caboose. So before the engine gets back a little crowd gathers and shouts "good-bye" as the train pulls out.

Imagination assured us that we should go puffing into camp with the sound of axes ringing on every side; the crash of falling trees echoing from the distance and drivers shouting as they urged overburdened cattle along the road. Reality stopped the train on a level stretch of track where a branch road joined and a few houses showed in the distance, but there were

KITSILANO CAMP, British Columbia, 1890. Logs were yarded to railhead by yoke of 16 oxen, log train dumping them on shore of English Bay, Vancouver, at foot of Trutch St. Rolled into salt water they were boomed and towed to Hastings Sawmill. (Vanc. City Arch. photo).

no indications of a camp to inexperienced eyes.

We were at the spur, and as the train was going on to Granite before running up to the landing, the only thing to be done was to jump off into the snow at the side of the track and walk a mile over ties to the collection of shacks which comprise the camp, with the exception of the neat little cottages of Mr. E. J. Lane, manaager, and Mr. George Rofe, bookkeeper. These are built on an elevation a quarter of a mile from the camp proper, which consists of the commissariat, the store, fifty feet in length, the cook house and dining room, perhaps ten feet longer than the store and wide in proportion, the five bunk houses with accommodations each for twenty-five men, the meat house, oil house, blacksmith shop and stables.

In addition to these buildings are tents and shacks where live the better class of men, who through necessity of working in the woods, do not care to mix with the genus lumberman of the bunkhouses. In addition to these are the cabins of a number of Scandinavian teamsters whose families are with them in camp.

Special sales are not in the line of the wood camp store. It offers novelties however and these consist of red, green, plaid, and striped mackinaws, boots with soles like iron and laces like ropes, leather belts mistaken by us for parts of a horse's harness as the iron buckles seemed too heavy for men's wear.

In the hardware line the novelties consist of wicked looking double-headed axes, saws with teeth that might gnash, springboard irons, crowbars, pickaxes and iron wedges of various makes, with a murderous looking array of iron-tipped peaveys. There is, as a matter of course, a grocery side which is perhaps of the greatest importance with between two and three hundred hungry men to feed.

At the back of the building genial Mr. Rofe has his office and makes such entries in his books as: "Twenty-five tons of flour consumed by the men in camp during the year 1900." As the nearest physician is at Granite, ten miles away, Mr. Rofe has fallen into the habit of dressing the men's cuts and sprains when they are serious. On the rough shelves at one end of the office is an array of jugs and bottles that would make a city drug clerk open his eyes with dismay. Vaseline in gallon cans, painkiller in two-gallon jugs, court plaster by the yard and absorbent cotton by the bale. The wood-chopper takes to medication as a duck to water, and it has been necessary to have a generous supply in camp.

There is a window cut in the rough board partition dividing the "office" from the storeroom; in this window is a circular revolving shelf just large enough to hold the great ledger in which the woodsmen write their names — those who are scholarly enough for this feat. It must be said that, though one occasional makes "his mark", there are names, real or otherwise, frequently inscribed on the pages in faultless "copperplate".

The revolving shelf was devised by Mr. Rofe to prevent the ledger from being torn to pieces by the men dragging it back and forth in their anxiety to see for themselves the amount received during the month, the rules for paying them being elastic, a necessity if the camp is short of hands, for if refused their wages the moment they are earned the men are liable to take offense and leave.

The names on the ledger show the cosmopolitan character of the camp, most nationalities being represented, from Russian Finns to Maoris from New Zealand, though perhaps Scandinavians predominate. Out of forty names beginning with J, twenty-one are John Johnsons. The Maoris came to camp last year with an Englishman from Australia, addressed as "Old Man Stewart". The trio had been to the Klondike, and meeting with reverses, came down to Seattle and drifted to the wood camp. The New Zealand men did not object to work but they did object most decidedly to chopping with the double-bitted axes that the white man used so effectively.

They were convinced the extra blade would be driven into their bodies in the backward swing, and the manager was obliged to order single-bitted axes for their use. One of these islanders aspired to an English name and was called Jesse King; the other was satisfied with his aboriginal name of Manara.

Another character was a big teamster with a saber cut across his face who was known as "Mulvaney", and was in personality a duplicate of Kipling's hero. Like him too, he had seen service in India. Other soldiers who had fought under Kitchener in the Soudan have served a term as teamster or wood chopper in the forests of Washington, and have taken pride in showing their credentials to this claim in the idle hours of the evening to their bunkhouse mates.

There is another class of men besides the foreign ex-soldiers who take refuge here. One evening three men walked into camp over the ties, a necessary mode of travel with most of them. Two were wood choppers unmistakably, but the third was looked upon as a tourist seeking a sensation or a reporter working on an assignment. He, however, handed out an employment agency ticket with a hand white and guiltless of marks of labor, and the ticket showed that William James had been engaged as a "skid greaser". The name was inscribed on the ledger immediately under that of "Clarence Montaigne", teamster, and it afterward proved that Mr. James, who could grease skids very well indeed, was a lawyer from an Eastern city, the son of a prominent New York citizen and son-in-law of an Eastern college professor, well known. The young man was, owing to troubles of a domestic nature, seeking isolation from his world for a season.

It requires a man of steady nerves and cool head to fill the position Mr. Rofe occupies at camp; for desperate characters find their way there, and when under the influence of liquor easily take offense and don't hesitate to threaten annihilation to anyone who crosses their path. One day last winter a question of the ownership of some tools arose, and the man who claimed them accused the bookkeeper of favoring the other. In Mr. Rofe's own words: "Things looked pretty black when the man came at me with a ten-pound iron-banded mallet". His coolness however, saved his life, as it has done on other occasions. His experiences are sometimes as amusing as startling. He is mistaken by the erratic chopper for some former associate, and

CAMP BARBER does Sunday morning chores at Camp 5, Chemainus, B.C. as other loggers mutter, "Do they think we're some kind of animals, wanting pictures of us?" (B.C. Prov. Arch. photo).

mysterious passwords are whispered in his ear and significant grips given his unwilling hand.

Meal time at camp is the hour of interest to a visitor as well as chopper. In the dining room are five long tables with benches on either side. Plates for twenty men are laid at each table. The cutlery is substantial, the dishes solid and the provisions plentiful. The room is warmed by a great iron box set in sand and capable of swallowing half a dozen sticks of cordwood. The adornment of the walls is not considered necessary in camp dining rooms; there is however, one picture in this. It might have been placed there to frighten the men into curbing their appetites, or merely as a reminder of what they may come to in a period of desperate idleness. It is the face of a bloodthirsty villian with a revolver. Whichever way you turn the eyes follow you and the revolver is pointed at you with startling persistence. The quantities of food that met our eyes as we peeped into the

kitchen just before the noon dinner proved that this realistic picture doesn't in the least affect the men's appetites. A huge roasting pan was filled with great chunks of beef — the very best cuts, for whole beeves hang in the meat house, and the head cook chooses for himself. Five gallon kettles of soup, of tomatoes and other vegetables were simmering on the range which filled one side of the shed. The young cook — Joe Reilly — who looked more like a soldier than a chef, was dipping up tapioca pudding with a young shovel into dishes the size of individual tea trays, and in an enormous wash tub bread for the daily baking was raising.

"Anything at all", won't do for the choppers at this camp. They know what good living is, or they think they do, which amounts to the same thing in making the one presiding at the stove uncomfortable. One Sunday morning Joe planned to give them some especially nice cuts of steak. An epicure would have smacked his

EVENING REPOSE in Heriot Bay, B.C. bunkhouse in 1873. Said one logger, "We tied our wet clothes to wires so they wouldn't crawl away. (Vanc. City Arch. photo).

lips both at the quality and mode of serving; but the man at that first table just couldn't understand the privilege of tasting the juicy sirloins and porterhouse. "Take this back to Joe," he said to one waiter, plunging his steel fork into a tender bit, "and tell him to cook it until its fit for a man to eat. Does he think we don't know dog meat in here?"

Whether from force of example or some other cause, the appetites of the men in the dining room are similar; for a certain period they ask for nothing but the plainest food, bread and meat being the staples; then a sweet wave strikes the dining room, then enormous quantities of pies, puddings, cake and sweet sauces are consumed. And when their orders for the change are given, whether they conflict with the cook's convenience or not, they have to be filled to the letter or there's trouble for the management to settle.

One studying types has opportunity to do so

as, at the sound of the triangle, the men troop in to their meals. Here are faces from all parts of the world, most of them totally unlike in features and habitual expressions. Yet as they go swarming through the door intent on satisfying their hunger, a similarity of appearance strikes one. It may be their dress — the mackinaw or overalls of the woodsman — but that can hardly affect their faces, and on the majority we read the aimless, ambitionless look of those who are content to be one of a class. Eating with them seems to be a task to be gotten over with as soon as possible. They haven't even time to allow the waiters — there are four in the dining room. — to bring their coffee, and the great pots for coffee and tea are set on the table before the men are called. They pay no attention to the serving of courses. The pie is on the table to be eaten, and if so inclined they begin with pie, and if soup strikes their fancy later on they end with soup.

63

Out in the woods they are at their best. When the storm cleared away on the day after our arrival, work was resumed in camp, and the sleds, twenty feet long and drawn by fine teams of horses, went by ceaselessly, the men singing perhaps, and in many cases were surprised at the magnificent voices of these teamsters.

Felling the trees gives them opportunity to display their great strength, which in some cases is accompanied by a certain native grace wasted on the wilderness. They stand on springboards whose irons are planted in the great trunks eight or ten feet from the ground, as the base of the trees are wind shaken and both difficult to split and useless for lumber.

Chop, chop go the men with their great double-headed axes, then the teeth of a long cross cut saw tears through the wood, and the tree begins to tremble, then starts downward in the direction that the "undercut" indicates it shall go, and the choppers raise their warning cry of "Timber!"

From the window of Mrs. Rofe's cozy sitting room it is interesting and exciting to watch one of the great firs or cedars go crashing downward, striking the trees in its path and bringing the smaller and more supple to earth with it, and then observe the standing tree when released spring quivering back to its place. But out in the forest where the stranger to sylvan sounds is uncertain of their direction, the knowledge that a tree is about to fall brings nothing but terror lest it fall upon him. Accidents from falling trees do sometimes occur, even to the men accustomed to the woods, but there is always danger from the rebounding trees when the choppers are not on their guard.

There is one necessary piece of work that at the present time is done by boys exclusively — that of skid greasing. With a floppy rag at the end of a pole they go about dipping the rag in a can of crude petroleum and slapping it upon the skids or hewn logs that form beds for the sleds to slip along on. Great barrels of the disagreeable stuff stand at convenient distances along the roads leading to the wood piles, and the "greasers" may be seen, when not at work, warming themselves by fires built at the side of a huge stump. Levi Gordon, a "greaser" of perhaps 14, assured us that he liked the work. "Besides", he said, "some one has to do it, or they couldn't haul wood over the skids at all. I go to school at Getchell, two and a half miles from here, part of the time, but I like skid greasing just as well, and besides I am earning money."

The teamsters mostly own their own horses and haul for a certain amount per cord. If inclined to be tricky, and not a few are, they make the task of measuring the wood a most disagreeable one for the manager. The wood is unloaded from long, narrow sleds, and piled at the landing, a platform from which the cars are loaded and where the wood is measured. This piling can either be done compactly, or the sticks beneath so criss-crossed that one cord can be stretched in bulk well toward two, and not a little trouble arises over this cheating in measures.

While the majority of teamsters work by the cord, the choppers mostly are "day men", a system of employment perhaps quite as satisfactory to the company owning the camp.

The wood cut in one week here averages 900 cords, and the men can, if so inclined, make good wages, though the greater number have no idea of saving, and spend any sum they may accumulate in drink as soon as the fit strikes them to leave camp and go to "town". However there are a few of the younger men who are in the woods for the purpose of earning money to finish interrupted educations or to enter some profession. In Woodland camp at the present time are two such cases, the young men living in a tent, studying during the spare hours of evening, and having little in common with the men by whom they are surrounded.

The Northwest Lumberman's Evangelical Society has recently sent missionaries to this camp, as to others in the West, and they are hopeful that the men may be benefitted by their instruction as to higher channels of thought and life.

"CROSSHAULING THE COOK", as early loggers termed "beating the gut hammer", ganging up for dinner in an Oregon camp. Below, tables ready for the rush. Yet no matter how ample the meal the cook expected growls. (Ore. Hist. Soc. photos).

TWO MILLION TIES cut from lodge pole
pine at Hawkins Creek, Yahk, B.C. (B.C.
For. Serv. photo).

Bullcook Blues

Somewhere the sun is shining, somewhere hearts are gay — but not at five o'clock in the morning in an Oregon logging camp when the bullcook gropes in the blackness of his shack for a billet of stove wood to smash the alarm clock. Nothing else in the camp stirs. No sounds break the abysmal silence. There is no sign within six kilometers of the donkey engine to indicate anyone else in the place is even alive.

So fully believing everybody is either dead or the camp closed down during the night to leave him and the cat to watch things until the new owner takes over, the bullcook packs in a lip full of snuff, pulls his pants on over his underwear which he never sees in daylight so doesn't know it is three weeks past the time when he thought he ought to wash it, snaps his red suspenders which pulls a safety pin loose, and goes to work.

With his little can of kerosene he paws his way through the chilly dank and starts fires in all the shacks with the greenest and wettest woods the push can find. It may sometimes be necessary to melt the stove pipe, set the roof ablaze and even the shack itself, but if the bull-cook has no higher function in life than to start fires, he always makes them burn.

This applies to the bath house heater too, after which he goes to the cookhouse for some coffee, providing he is allowed in this holy of holies. The hot brew gives him the feeling that he may stay on in camp until payday. Having made this profound decision, one he has been making every morning for six months, he sniffs the air at the door, watches three crows in the high branches of some hemlocks and observes with Socratic solemnity that the day will be fair and warm if it doesn't rain or turn cold.

Now to fill the lamps, and with a fine show of pride clean them. He lines them in a row on the floor by the stove. He grabs a chimney, shoves it once into a pail of water and sets it upright on top of the still warm stove. He does this with each and the result is astounding. The steam from the water inside the chimney cleans the glass and leaves it with a gloss that would shame crystal. For the forty-sixth time he considers the possibility of getting the process patented.

The sweeping of the shacks is done with more cupidity, the net take sometimes as high as thirty cents a day from the poker and casino leavings. It all depends upon how catty he is with the Baldwin Retriever. This is a combination spoon and claw with which after long practice he can pull even a sticky Chinese yen up through a crack or from under a split board.

Bucking wood for the cookhouse is a dull enough afternoon pastime lightened only by the forlorn hope that the wood assigned him is dry enough to get a grunt of acknowledgement from the cook. After that he sneaks an hour in the hay before rebuilding all the stove fires and relighting all the lamps. After supper he does not join in any money games but spends an hour sniffing out a bottle of bourbon and scrounging a drink. Thus warmed, with a happy glow in his stomach he returns to his harem where magazine covers of pretty movie queen pictures look down at him from the walls of his bunk. He takes out his fourteen-ounce silver watch, winds it and hangs it on a nail.

It is ten o'clock and he gloats on the thought that there are three more of them until pay day and a ride down the river. Three days more, he says to the half-dressed damosel on the beer calendar, and blows out the lamp. Three more days. Dreamy thoughts carry him along on soft, billowy clouds. Blonde . . . red-head . . brun- . . .

Article reprinted from "The Sad Life of a Bullcook" by Otis Beale in Loggerheads, October, 1922.

BIG LOG, BIG TREES in Washington's big woods. Skidroads for oxen and horse skidding were built with almost as much precision as railroad track grade; each log must take skid without bumping and in rounding curves must keep to center of road. (Uni. of Wash. photo).

And TIMBER became a CROP

It was noon on Baldy Ridge, a high windy area above Clear Lake, Washington, where Side 4 was logging against time. Nels, the fireman, had boiled his coffee and the crew, sprawled on the shady side of the unit, gulped down the strong brew from tobacco cans and dug into their "feed bag" lunches brought from camp.

Sunlight poured down out of a hot blue sky, glancing from muscular, sun-browned torsos and hard faces grimed with the sweat of a morning's work, prodding the heat devils into an erratic dance over the tangled miles of slashing.

"We'll finish her this afternoon," said Big George, the hooker.

He pawed up his gloves and went out to the haulback bight to plot subtle wizardry with steel line and tail block; presently, the nooning over, chokerman and whistle punk joined him. The big whistle bellowed its long and short, the little whistle answered with a snappy "skin 'er back" and the game was on again.

About 3 the last turn was shaping up. Chokermen bounced in, quick and sure in their caulked boots, gloved hands reaching for line and hook. The logs were noosed, chokermen ran for the clear and Wally, the rigging slinger, barked his staccato "Hit 'er" that meant highball from Blaine to Crater Lake. On the unit, the engineer opened the throttle and the turn came smashing to the landing.

Big George, two-twenty and as light on his feet as a cougar, balanced himself on a down sapling. And in a gesture unconsciously dramatic, he waved to the alert signalman who stood waiting, coil of whistle wire around his neck, electric bug in his hand.

"That's all," said George, the hooker. "Get in your wire, punk."

The last turn was in from Baldy Ridge.

It was not always like that. At the turn of the century and before, the woods were quieter. Axes beat out a steady tune, punctuated by the occasional crash and thump of a falling tree. A skid road defiled the forest, stretching down the green vista. Far down the road there would come a distant stir, an indeterminate sound, growing in volume to finally resolve itself into several elements — a scrambling of shod feet upon the logs, an incantation of lurid oaths rising to frenzied shouts and dying away into a grumble and the bumping progress of a heavy log.

From time to time the cavalcade would snatch a minute's rest and the six or eight pair of oxen would ease their necks in the yokes and puff a steaming cloud from their nostrils. A minute's rest, then up would spring the bull driver, red-faced, red-shirted, pouring forth a stream of blasphemy that would move the pyramids, whacking the "gad", prodding and tongue-lashing the oxen with professional skill and pride.

"Gee, haw. Star! Muley! Hey Boy. Blankety-blank, blank blank!" the bullskinner would shout until he was tired of the sound of his own voice. Then the scramble of hoofs, the heaving of several tons of live muscle against the dead weight of 15,000 board feet of raw timber, the slow and slipping progress to the beach or bluff's edge where a long chute might receive the cut and drop it expeditiously into the waiting sound or lake.

The bull driver was the king pin of the lumber camp. He was on top of the transportation problem. He got the biggest pay of them all and his duty and satisfaction was to keep the cut moving as fast as ox flesh could move it between stump and log boom. It was a matter of dollars and business sense.

So was the early logger's selection of a skid road. It must be all down hill, the width generally 10 feet, sometimes 12. Skids of fir about 10 or 12 inches through were embedded to half their diameter in the ground, about 9 feet apart so a log 20 feet long would have a bearing on two skids and be in no danger of upping and burying its nose in the ground.

Next came the landing or rollway. When a suitable bank was found a stout skid 20 inches through and 100 to 120 feet long was buried lengthwise at the edge of the bank level with the ground, this called the brow skid. A number of smaller skids were pinned to it, following the incline of the bank to prevent the logs wearing the bank away and undermining the brow skid. A pitch of about ½ inch to the foot was given the bed skids to facilitate the rolling of the logs.

And back there were the choppers, working in pairs. After selecting the ground for the tree to fall on they cut a notch 6 inches deep as high above the ground as they could reach. In this was inserted a springboard, which was merely a 6 foot plank, 6 inches wide with an iron shoe to prevent its slipping. Often enough the tree was cut off as much as 10 or 15 feet above the ground. An undercut was chopped out, the tree falling in this direction, the saw cut made on the back side, wedges driven in the saw cut to keep the saw from binding and pitch the tree forward. The amount of pitch in the lower trunk often necessitated a liberal use of kerosene on the saw, the chopper hanging over it an inverted bottle with a small hole in the cork, so a few drops of the oil would fall on it at every stroke.

Logs were generally cut in 24, 32 and 40 foot lengths which were standards for the sawmills. When a tree was felled, trimmed and bucked to length, the swampers raised and flattened one side of it and beveled off one end so it would not catch on the skids and plow them out of the ground. When the hook tender with a pair of oxen got it and others yarded out, they were coupled together ready for the main haul.

The skid greaser with his swab and can of grease took his place in front of the lead oxen, swabbing each skid before the passage of the train of logs. On arrival at the landing the chains were uncoupled, logs scaled and with a groan the big sticks were rolled to their final resting place in the water.

All this changed little through later years except in the effort to get the logs out of the woods off the ground. But the day of the grizzled bullpuncher was fading out. The last of the breed may well have glared with contempt at the snorting, barking, straining of another animal, this one of steam and steel, the donkey. Yet the "vertical gypsy" and improved models doomed the bull driver and swept his cud-chewing charges clear out of the woods. Heavy teams of horses replaced oxen in some areas for a year or two but in the late 1880s only the line horse remained.

In that day loggers sat on the edges of their bunks and swapped yarns about the sagacity of these four-footed timber beasts. Some of them became legendary heroes. The fact was they were an outcome of the necessity of getting the gypsy line returned to another log as speedily as possible. The horse, without blinders, without harness other than collar and a pair of traces hooked to a single-tree, performed this task for a short period until the cost of his care and the swamping that had to be done to enable him to work at all, spurred logging brains to figure out how to replace him. While the horse was on the logging camp roster, however, he paid his way with plaudits. Without lead or halter, at the word of his driver, he came and went, made his pull and footed it out of danger.

While an amazing amount of board feet was snaked to the landing by this method, very big timber in rough places was a problem. Some of it had to be left and some cut into logs short enough to be handled. Yet old time hookers knew holds that increased the power of the little engines astonishingly. There was the "luff", the "luff with a whip," the "block and a half" and the "two block" purchase, rigged on fractious logs by clever arrangements of line and sheaves, choker and grabs. Also in the '80s logging was mostly on level or gently sloping land. With pole-chutes and skid roads as a further aid to hauling and with plenty of grease on hand, plus more patient men and an 11-hour day, logging was successful according to the times.

BULL TEAM in Oregon woods — 10 oxen
with bull at right on third yoke from front,
bull puncher with goad first man sitting on
rump. (Ore. Hist. Soc. photo).

(Left, opposite) NO CROSS LOGS on skid road here, on Big Sandy, Oregon, in 1889, as barked logs slid easily down steep grade. Logs were usually coupled together with chisel-shaped hook on either end, called dogs, extra chains and gear carried on narrow sled called PF. Note fallers on spring boards rigged on big hemlock in center of photo, hats hanging on snags of branches. Opposite page bottom, another ox team set in southern Oregon. (Ore. Hist. Soc. photos).

THIS WAS LOGGING (below) near Klamath Falls, Ore. in early '70s. Wheels of wagon were made of crude concrete and sawdust packed into iron tires. (Ore. State Library Arch.).

BRITISH COLUMBIA skid roads in late
80's. (Above) Brunette Sawmills' 16-ox team
moves logs to landing on Nicomekl River.
Opposite page top, Fraser's logging camp
and bottom, note method of chaining logs
together. (B.C. Prov. Arch. photos).

Genus Homo: Habitat Deep Woods

Back there in the tall uncut lived and worked a legendary man-animal who will go down in the mists of the valley just as clearly as St. George who slew the dragon and other saints and sinners of the story books — the logger of tradition. He was of a breed hell-bent to chop, saw, drag and slash himself right out of existence.

He would deny this. Loggers, from the lords of a hundred open-handed camps to the bullcook of the meanest, most haywire layout south of Skookum Creek were unaware that off there in the dim and distant was the last of the big trees, the bottom of the chute down which the final ride was heading for the big splash. Or else they just didn't give a damn.

From Penobscot spruce, Bangor and booze, bawds and battle, to Saginaw pine, Columbia River fir and Chinook squaws, there had always been timber, timber, timber without end and the thing was to cut it and live it up. The logger had only to realize he was on the run west with civilization, farmers, cities following him like a black memory. He had only to look at the growth rings in his logs to figure timber didn't grow faster than saws could cut it. But who cared for growth rings? Muscles before brains. The hosts of Persia fell before the small organized phalanx but the logger said, "Look, son, up there by Lyman, over on the peninsula at Pysht, down in Oregon by Toledo and K Falls— timber 'til hell won't have it. Us loggers and our race will live forever." So they will — in history.

A likeable stalwart, the logger. He was careless, improvident, wasteful — a man of to-day, not tomorrow. He was honest, especially when the camp was secure. He was helpful to others, having sense enough to know he might need help himself anytime. He was generous because he lived in nature's heart where there was no limitation, no end of logs or pay days. And why worry when he could always end up

his days where he began, on the skid road?. He was rough in manner, a caustic critic with small respect for authority unless it had genuine ability behind it. He was kind at heart and tolerant of mistakes born of inadvertance or ignorance.

Out of his natural habitat, like other strong characters in strange surroundings, he felt self-conscious and tried to off-set this by some braggadocio. Under stress which required non-chalance he might as well have had one hand full of fir needles, the other clutching a bouquet of ferns. He distrusted towns and yet they had an unholy attraction for him. He had a good-natured contempt for "paper collar stiffs", tolerated sawmill workers but fought shingle weavers on sight and helped maintain police court costs.

Being guileless, the logger was preyed upon while in his cups by booze heads, harpies and crooked bartenders but up to that stage he was clam-happy in any saloon or dance hall. Surrounded by his woods mates, it was his profound privilege to sound out, "Drink up, you stumble bums. I'm buying!" And somewhere along this happy rollway, truth and cold facts fell under the bar rail and feats of fancy spilled out until the air was filled with the glory of logging that never was until an imaginary "side push" reached in and laid bare knuckles on his bad eye and he did a quick fade out on the sawdust floor.

Loggers of the old days were not all like this but others were a lot more so. Some were silent martyrs, some superstitious, some could quote Shakespeare. To a man however they knew the deep woods would be their home forever. They could not foresee the day when some mother would no longer say to her daughter at a country dance, "Now Annabel, you be careful with them loggers. When they're dressed up you can hardly tell them from men."

BULL TEAMS on haul back to woods on
Thurlow Island, B.C., 1900. Only finest of
timber was cut, this rarely more than mile
from river or "salt chuck". (B.C. Prov.
Arch. photo).

They Had to Shoe the Oxen

The logging camp blacksmith was a man for all seasons and busier than a hungry cowbird, depending on the size of the operation and amount of his pay. In a big outfit using twenty to thirty oxen there would be two teams of them, one for getting logs out of the heavy cover to the skid road, the other for dragging them to the beach or river.

Either team would use one or more bulls. Never were they yoked in the lead or on the butt but somewhere in between. They were always balky or mean and never willing to respond individually to the goad or blood-curdling roars of the bull puncher. But they could pull their share of the load when the other animals moved.

And they all had to be shod. The cloven hoofs had very thin shells which was fine on the farm where they walked on soft earth, but for tramping over broken timber and hammering down on cross logs and rocks on the skid road, the hoofs had to be shod with iron.

The easiest way to get the job done was to build a stall-like frame near the log dump where the oxen were unhitched, using light timbers fastened together with mortise and tenon and wooden pegs. One end was left open, a wooden stanchion fixed at the other, and a windlass set up on one side.

The ox was led into the frame, head secured in the stanchion. A heavy canvas or ox hide belt, attached to one side of the stall, was passed under the animal's belly, fastened to the windlass which was then wound up, lifting the ox off his feet so he could be shod. In the small outfits which could not use or pay for a full-time blacksmith, the second teamster with a man helping him might do the work.

SHOEING OXEN in Oregon. (Ore. Hist.
Soc. photo).

Loggers and Log-heads

Out yonder the hills are stripped and bare. Ugly brown and black patches foul the slopes like fungus growths between the green areas of young forest trees. Devil-machines roar and clatter and yank little logs out of the thin timber and pile them like matchsticks on giant trucks. And the night wind sighs over the ghosts of old bunkhouses and the mighty men who lived in them with small convention or conformity.

The early logger was his own man. Generally he went along with the rules because it was less trouble. But in every camp there was a generous sifting of rebels, reactionaries and the off-beat characters who colored the long days with the swish of a gaudy brush. The isolation and freedom of the big woods camps attracted and nurtured those rugged eccentrics who groped their way through life by dead reckoning and spiced it up for the rest of the timber stiffs.

In the 4 *L Lumber News* of January, 1927, an anonymous writer recalled several loggers with pronounced quirks of character "The first to come to mind," he wrote, "I will call The Silent Man." I never knew his name, nor did anyone about camp unless it was the timekeeper. He was with us about two weeks, as I remember, and during that time never spoke but once. He would rise, dress, wash, go to the cookhouse to eat, and thence to work. At noon he would come in, wash, eat, and return to his job. At night, again to washhouse, cookhouse, and back to his place on the deacon seat. Here he would sit for an hour or so looking about him in silence and then retire. He never read any of the scanty periodicals about camp, never wrote a letter, never exchanged a word with anyone during the fortnight. If anyone said, "Good morning," even, the salutation was met with a bleak silence. Finally, one night one of the men made bold to ask him whether he had long been on the Coast. Came the reply, "My friend, I have been on the Coast a good many years, but I want you to understand that I am not going to undergo any cross-examination." The next morning he quit.

The "side-push" afterward told me that the silent one had upon his departure remarked, "They want to talk too much in that shack." Some thought that he had a past. Maybe. My theory is that he saw no advantage in forming acquaintanceships and that the conversation was of no interest to him. The camp was new to me then; later on I got his viewpoint—if that was it.

Like the silent one, many of these queer characters did not stay long. There was the skidroad man who worked a day and quit because he couldn't have grade stakes to shovel by. There was the chokerman who was with us three days or so and who, one evening as we passed the blacksmith shop, called our attention, with a fervent appeal to the second person of the Trinity, to the fact that the grindstone wobbled on its axis. We laughed. Next morning he rolled his blankets and left, remarking, "I never did have no luck in a camp with a crooked grindstone."

Many and diverse were their superstitions. I recall a strong, elderly man who worked with the riggers' crew — for we had one of the first overhead skidders on the Coast. The head rigger, a methodical chap, allotted a part of the tools and equipment to each of his men to carry, and each was responsible for his part. It chanced that a coil of rope used to haul rigging blocks, etc., up the spar trees, fell to the lot of this old fellow. Carry that rope? I should say not! Nothwithstanding that it was his first day in camp and that, patently, he was not well supplied with funds, he refused pointblank. He would carry anything else. But that rope — no sir! "But why?" queried the head rigger. "Well, you see, one of my ancestors was hanged," replied the beginner, "and I think it's bad luck for me to pack that."

We laughed. We hooted. We intimated that we thought him lazy. But all to no purpose. A change had to be made. We found later that he would willingly handle the rope, work with it, coil it — anything but carry it. The theory of indolence, we found, wouldn't hold water. I've seen him staggering up a mile-long hill in the hot sun with the sheave from a 200-pound tailblock on his back, while another man strode ahead with the comparatively light rope carried like a bandolier over one shoulder and under the opposite arm. Truly, the ways of the human mind are manifold and strange. . . . .

Our perennial bullcook. He had a poker "system" which he employed as a manual of tactics in continually laying siege to Dame Fortune. His faith in his system was both sublime and pathetic. Bullcooks do not draw big wages and this one was always broke within three days after payday. For a while his spirits would droop, but long before the next payday he would be planning the renewal of the fray. When I left the woods he had not been out for nine years. As I shook his hand in farewell he cried, "Some day I'll make it big and then I'll go out with the boys." I didn't have the heart to smile. All I could say was, "You're bound to make it sometime, Jake."

One of our donkey punchers was the bullcook turned wrong-side out. He would act the miser for six months to play the spendthrift in town for a day. Laboriously, he would patch and mend castoff apparel, deny himself the meager luxuries of the commissary, pinch and scrimp, and grumble at the cost of board while counting the days to a shutdown. Then he would sally forth with great expectation, only to fall among thieves and be stripped of everything but his raiment while sodden in the back room of some bar. When he came to himself he would sadly return to camp. The classic greeting was, "Well, old boy, back for six months more at hard labor, eh?" This catastrophe occurred regularly twice a year, so regularly, in fact, as to lend color to the theory that the banditti lay in wait for our engineer and viewed his approach from afar.

Not all of these eccentrics were so constant to one camp. There comes to mind the rigging slinger, Slim, a boilermaker by trade, who boasted that he knew every town in the United States. Skeptical, I murmured, "Watertown, New York." "Well," he replied reflectively, "I wasn't there very long, but I worked in the brake shop and boarded at the Crystal Spring hotel on Factory street." "Holy smoke!" I gasped, "you win." The place was an obscure upstate town near which I was born and the description was exact. Further tests by others failed to shake the reputation which he had given himself. While Slim should not, perhaps, be classified as eccentric, I feel that he was at least unusual in this respect.

More unusual still was the wanderer we called "Paddy." Paddy had a boy with him whom he was teaching "to be a good tramp," he said. He delighted to work barefoot in the snow as a sign that he was a "tough man." Two very odd ambitions, it seemed to me. If asked whence he came he would always answer, "From all over hell, and I'll be *from here* if the board don't get better."

And among those from far places was the Irish lad from "just across the river Kerry, a little way from Clare." The fact that I had read all the way from Kant's *Critique of Pure Reason* to *Old King Brady* and the *James Boys* impressed him not. But the race is not to the swift or the battle to the strong. One night I chanced to recite part of D'Arcy's "Face on the Barroom Floor," and presto—my reputation for erudition was established. This rhyme was his ideal by which all poetry was measured. He had a similar definite stand in music. The one bright spot in his life seemed to be the time when he had heard Chauncy Olcott sing, "Where the River Shannon Flows."

LOG ROLLS INTO BURRARD INLET, Vancouver, B.C., from hillside chute. (Vanc. City Ach. photo).

WASHINGTON SKID ROAD passed ox barn, no more than shed roofed with hand-split shakes. Note method of chaining yokes together.

Special tallow was used to grease skids, boy of 17 or 18 carrying can with swabbed stick, running ahead of lead log and slapping grease on each cross log. (Uni. of Wash. photos).

(Opposite) FIRST WHEELS in Washington timber — both oxen and skid road get warning of future uselessness. (Uni. of Wash. photos).

HORSES TAKE OVER skid roads in early 1890's — this team in British Columbia. Teamsters were as tough and vituperative as bull skinners, one here accompanied by dog. (B.C. Prov. Arch. photo).

(Left and opposite bottom) HORSE TEAMS in British Columbia. Bottom photo shows loggers riding to camp on last turn of logs at McKay and Flannigan's, Hazelmere, B.C. (B.C. Prov. Arch. photos).

(Below) 10-HORSE TEAM at Park Kells District, B.C., 1891. (B.C. Prov. Arch. photo).

FATEFUL DAY for "hay burners" on this
Surrey, B.C. skid road as horses were replaced by donkey engine after this photo
was taken. (Vanc. City Arch. photo).

(Opposite top) BIG WHEELS were first
effort to skid logs off ground, major logging
development. Forward ends of logs were
lifted under axles, rear ends dragging.
Scene here is at Crows Nest Pass Logging
Co. near Echo Lake, B.C., in 1921. (B.C.
For. Serv. photo).

(Opposite bottom) TWO-SPAN TEAM
snakes log down skid road half-mile from
boom — Hilton Logging Co. Oyster River,
B.C. (B.C. For. Serv. photo).

(Opposite top) VERTICAL SPOOL DON-
KEY, one of first used in British Columbia.
Bottom, Mud Bay camp, Royal City Planing
Mills, Hazelmere, B.C., 1898. Celebrated
locomotive "Old Curley" is in distance.
(Vanc. City Arch. photo).

DONKEY NEEDED HORSE. Line horse
waits to take gypsy line to chokermen for
another turn of logs, Camp 2, Stocking Lake,
near Ladysmith, B.C. in 1902. (B.C. Prov.
Arch. photo).

3081

FALLERS MAKE UNDERCUT on big fir
in Oregon, working on springboards in
square notches to get into softer wood above
spread-out base. Opposite, mammoth spruce
near Toledo, Ore. will fall toward camera
after saw cut is made on back side. (Ore.
Hist. Soc. photos).

EIGHT-FOOT DOUGLAS FIR succumbs
to axes in British Columbia forest. Springboards were usually iron-tipped to keep
them from slipping out. (B.C. Prov. Arch.
photo). Opposite, bearded and burly Oregon
faller used one springboard to stand on
while securing another one higher and
cutting "scarf" for undercut. Tree was
perhaps "shaky" close to roots. (Ore. Hist.
Soc. photo).

WASHINGTON FALLERS start saw cut as woods boss looks on. Springboard was used even when cut was this low to ground. (Uni. of Wash. photo). Opposite, number of axe strokes could be counted by chip marks on undercut. Crosscut saws ranged in length from 8' to 12'. When necessary to make longer, section was brazed in middle. (Ore. Hist. Soc. photo).

OREGON BUCKER eyes cameraman wistfully as he pauses at job, toughest in woods, working 10 to 12 hours a day. (Ore. Hist. Soc. photo). Opposite, fine, stalwart spruces in Washington forest stand ready for execution, fallers appearing not too happy about it. (Uni. of Wash. photo).

They Moved the Logs by Hand

It was called "hand logging" as only axe, saw, peavy, brute strength and screw jacks could be used to get the logs into the water. This meant trees could be felled only on steep slopes where gravity helped the weight of the logs. And this in turn meant very little hand logging was done except on the islands and inlets of British Columbia where there were steep slopes and quiet water for holding booms which could easily be towed to the Vancouver mills.

It was man-killing and dangerous work. Unless a tree could fall and slide directly down to the water, which was a rare show, a bedding of small trees had to be laid crosswise so the log-tree could be run over them to a point where it would pitch downward on an unobstructed course.

Hand loggers usually worked in pairs, living in a shack whacked up in the area in which they had permits to cut timber. With the tree down they would start at the butt and carefully cut off the limbs close to the bark. Setting up a jack, either screw or Gilchrist pump jack, on one of the skids, they would slowly roll the upper limbs and knots up-side and dress them off as the other had been. If the tree was on a steep grade it might start down at any moment and the loggers would have to be catty to keep from getting caught by a knot or branch and dragged underneath to be crushed to pulp. Or the butt could swing on a wide arc or make a side swipe and bounce a man into a hospital or eternity.

If the tree did not start, a "ride" was barked and then the partners went down to the "nose" (top of tree) and slowly raised the end. Eventually a tree was off and running and a steel trap was never any swifter in action.

If a tree was running free, and it was not unusual for one to run half a mile, it could well be going with tremendous speed and force. They would often run through or under a tree much larger than they were and the smaller stuff did not fall at once, but seemed to stand for a space of time and then as though a big mowing machine had gone through, start falling in a swath from the running tree. This would hit the water like a meteor, completely disappear and then rear its full length out of water.

There were many tales told about hand loggers as they were most often loners or characters who could not fit themselves into deep woods society. In one case, two of them started a show using a hydraulic jack and immediately one got drunk and stayed that way for two days, his partner perplexed because he knew no whiskey was brought along with the grub stake. The sober one wanted to work but found the jack would not. He inspected the alcohol in it, found it had turned to water, and then he knew how his partner had kept drunk so long.

One set of hand loggers disagreed on how to cook cabbage and decided to split up. They smashed the stove, each took a saw and cut the boat in two, and threw all the other equipment into the salt chuck. Then they shook hands and went their separate ways. Not much reason to it all but then they were hand loggers.

HAND LOGGER at work near Tekearn Arm, B.C. Using only axe, saw, screw jack and "guts", the hand logger could wrestle out a living on the steep, timbered sides of British Columbia waterways. (B.C. For. Serv. photo).

FIRE
in the TIMBER

There was one word every early logger understood, fire, and one sound that was at once sinister and fearful, the ominous wail of the fire whistle. When he heard it the veteran logger could see the consequences all at once — smoke, flames, the ceaseless, gut-straining fight for control, the heat, exhausted men, confusion and "What the hell's it all for?"

In hot, dry fire weather the whistle pulled him up like a startled colt. He had no time to run and ask questions, how close the fire was or which way the wind was blowing. He left his tools where they were, ran over slash and down timber, even burning limbs dropping from high up, heading for the nearest point of security and feeling a deep relief to get there without being cut off by the blaze. Scores of men were coming from the timber in all directions, spreading out along the roads and trails toward the fire with shovels and mattocks or whatever implements they had laid hold of. When somebody pushed an ax at him he joined the rush on the double.

After days and nights of continuous struggle in burning brush and blinding smoke, when at last the flames were stayed and the camp itself made safe, the harassed, red-eyed, grimy-faced crew slept the clock around. When the forest cooled and the logger ventured back to his place of work he found his saw partly melted and his oil bottle a wad of fused glass. All over the woods were still smouldering remnants of what had been clean, sound logs ready for yarding in a scene of black ruin.

Early logging in the big woods of the Northwest caused a sudden and prolonged increase in forest fires but great areas of timber was destroyed when only Indians inhabited the fringes of it. The first settlers reported many small and large fires. The Hudson's Bay Co. factor at Nisqually made an item in his journal for August 14, 1835, "The country around us is all on fire and the smoke is so great that we are in a measure protected from the excessive heat". In 1858 the captain of a trading ship made mention in its log that the smoke was so thick off the Skagit River that he was forced to anchor and hundreds of suffocated birds fell into the water.

There was a fire in the Pacific Northwest woods in 1853 and a larger one in 1865 which made a charred sweep of all timber in its path. Both fires started north of the Columbia and Joe Matte, a French Canadian trapper employed by the American Fur Co., reported they were undoubtedly white men's fires as Indians seldom built camp fires in the dry summer season. But white men did and as the country filled up flames flared up to an alarming degree.

CROWN FIRE in British Columbia timber near Youbou. (B.C. For. Serv. photo).

TRAIL CREW cutting out burned timber,
Little North Fork, St. Joe River, Idaho,
after great fire in 1910. (Ore. Hist. Soc.
photo).

Catastrophe in the Coeur d'Alenes

The fires of Northern Idaho and Western Montana, which on Saturday and Sunday, August 20 and 21, 1910, burned together into one huge conflagration, burning over an area of more than a million acres, destroying approximately two billion feet of timber, 81 human lives, 30 horses, and thousands of fish, birds and wild animals, may well be classed as one of the world's greatest fires.

Northern Idaho had experienced one of the driest seasons known to the pioneers of that region; with the exception of a few light showers no rain had fallen since early in April and, instead of being calm as usual, high winds prevailed throughout the early summer, thus drying out the forests more than usual.

Forest fires began burning on the lower slopes early in May, which was two months in advance of the usual dry season, and their number rapidly increased through May and June, but the forest rangers with a small amount of help succeeded in extinguishing all of the fires up to the latter part of July. During July, however, the fires became so numerous that it was difficult to man them and watch them after they were placed under control. On the evening of July 26, a severe electric storm, unaccompanied by rain, passed over the Coeur d'Alene National Forest. The rangers and all others on the forest having anything to do with fires were immediately busy looking for lightning fires, and during the next three days after the storm fifty-two lightning fires were extinguished and many others discovered that could not be reached in time to keep them from spreading. By August 1 we had men working on twenty-two large fires on the Coeur d'Alene National Forest alone, and by August 15 we had slightly over 1800 men fighting these fires, including two companies of

W. G. Weigle, author of this article reprinted from 4 L Lumber News of August, 1927, was U.S. Forest Supervisor of Coeur d'Alene National Forest in 1910. That year he and his personnel battled forest fires 24 hours a day, including the "Big Blowup" which broke loose August 20.

Federal troops stationed near Wallace, with only a very small supply of equipment and pack horses at hand.

To feed and provide bedding for 1800 men, scattered for a distance of more than 100 miles along the slopes of rugged mountains, where there were no roads, few telephone lines and but few trails, and many of the crews being located from 25 to 60 miles from the nearest supply station, was more of a job than the uninitiated may think. There was very little sleeping done by the forest officers on the Coeur d'Alene National Forest during that terrible month of August, 1910. Pack horses were gathered in from all over the adjacent region. Supplies were concentrated in warehouses that were kept open all night so that pack trains would not need to wait. On account of the absence of telephones, runners were provided for all the big fires to keep the supervisor's office in touch with the character of the fires and needs of the crews. The railroads of the region gave us all the assistance they could, frequently stopping between stations to let runners and men off at advantageous places.

The strenuous fire situation on the Coeur d'Alene forest called for superhuman efforts, but by August 18 we had control lines around practically all of the big fires and conditions looked as though we would be able to hold them. And with the usual weather conditions we would have held them, but August 19 was a bad day with the fires breaking over the lines in many places. Such conditions at the present time would tell us that we had a very low humidity, but in 1910 we were not so familiar with the relation existing between low humidity and fires. The Weather Bureau records, however, show an exceptionally low humidity for August 19, 20 and 21, 1910, which explains the cause of the unusual fires at that time.

August 20 started out bad in the morning with a very heavy wind blowing toward the Northeast. This dry wind increased in velocity during the day, carrying burning bark for long distances, starting many new fires all over the forest. On the morning of August 20, the nearest fire was five miles in a straight line from the

RANGER HALM'S CREW at Timber Creek, headwaters of St. Joe River before Coeur d'Alene fire of August, 1910. (Ore. Hist. Soc. photo).

town of Wallace, yet the high wind carried burning tamarack bark over this distance and set awnings on fire in Wallace in numerous places. During the late afternoon of August 20 the dry wind reached the velocity of a hurricane, spreading all of the fires over the lines that had been placed around them, and starting new fires all over the adjacent country. The men had to abandon their camps and fire lines and try to protect their lives.

Although the wind was stimulated by the fires, it was not of local origin but crossed the wheat fields of Southern Washington and was heavily laden with fine particles of soil, so much so that it was easily visible and produced mud wherever it struck a wet surface. Its velocity increased during the late afternoon to the extent that it blew down or broke off thousands of acres of timber. This occurred immediately preceding the fire. There were thousands of acres where practically every tree was down and blown down so suddenly that the trees practically all lay in the same direction, or parallel to each other. Many large trees were picked up roots

and all and landed from 50 to 100 feet from where they originally stood.

The rangers in charge of the various crews were real woodsmen who knew their business and were well acquainted with the country. Had it not been so the loss of life would have been far greater than it was.

Ranger Pulaski, one of the best all-around woodsmen that I have ever met, had a crew of nearly 100 men stationed about 8 miles south of Wallace. On account of the fire spreading so rapidly, Pulaski, 40 of his men and two horses were separated from the remainder of the crew. Pulaski realized that they had to get out of that place quickly or be burned to death. Conditions looked favorable in the direction of Wallace, so they started for the town. They would have been able to have kept ahead of the fire following them, but they ran into a new fire that had started during the day.

This new fire cut off their avenue of escape toward Wallace. Pulaski being thoroughly acquainted with the country knew of a tunnel about 100 feet long on the "War Eagle" prospect, about three miles south of Wallace, which appeared to be the only way to safety. They thought if they could reach this tunnel they would be safe. They started for the tunnel and although they had to pass through the fire in numerous places, made possible by protecting their faces with wet coats and blankets, they finally reached the tunnel and the fire got there about the same time; in fact, one of the crew who had lagged behind was caught by the fire before he reached the tunnel and was burned to death. The tunnel had sufficient space for 40 men and two horses, but the terrific fire on the outside caused the cold air in the tunnel to rush out and the smoke and heated air to take its place, which suffocated five of the men.

After the men went into the tunnel, Pulaski tried to protect them from the smoke and heat by hanging wet blankets at the mouth of the tunnel but the terrific heat on the outside soon burned the blankets, and the timbers caught fire. The smoke rushing in nearly suffocated the men. Pulaski commanded the men to lie down in the tunnel with faces to the ground as there was less smoke on the floor of the tunnel than anywhere else. One man attempted to rush out of the tunnel which would have meant almost instant death. At the point of a revolver Pulaski commanded him to lie down, which he did. After the wet blankets at the mouth of the tunnel were burned Ranger Pulaski used his hat to get water from a small stream on the bottom of the tunnel which he threw on the burning timber, which retarded the fire sufficiently to prevent the ground falling in to close up the mouth of the tunnel. He continued to do this until he fell over exhausted, nearly suffocated, and badly burned. One man who had kept his face on the ground all the time regained consciousness and crawled out of the tunnel over what he thought to be the dead bodies of the other men, who were either dead or unconscious. The worst of the fire was over. He staggered into Wallace and reported the location of the men, and stated that he was the only one of the 40 who had survived.

He reached Wallace about 3 a. m. Wallace was then on fire. I immediately organized a rescue party who went out to the tunnel, the fire by this time having burned over the whole valley and died down, so that it was not difficult to reach the tunnel. Just as soon as the rescue party arrived they got the men out of the tunnel, many of them by this time having regained con-

sciousness. After getting out into the air they all regained consciousness except five, who were dead. Those who were not dead were taken in to the Wallace hospital where most of them recovered in a short time, while others remained in the hospital for several weeks. The two horses that had been taken into the tunnel were so badly burned that they were shot by the rescue party.

The remainder of Pulaski's men went up on a steep mountain peak above the heavy timber and were safe, where they watched below them one of the most spectacular displays of fire that ever took place.

Ranger Bell had a crew of 50 men on Big Creek, about 17 miles south of Wallace. When Bell learned that he and his men were entirely surrounded by the flaming forest he gathered his men in to the Beauchamp homestead, which consisted of a cabin and about one-half acre of cleared land in the midst of the heavy timber. While he knew this was a poor place, it was the best that could be had. Before the fire reached the little cabin a large white pine tree blew down, killing three of his men. The wall of roaring flames closed in on the little homestead, burning the cabin and burning to death the homesteader and seven of Bell's crew. The remainder of the crew got into the creek and lay with their faces buried in the water, while the hair burned off of the back of their heads and burned to a crisp the skin on their heads and necks. As their clothing caught on fire they wriggled into the shallow water and extinguished the fire. At the same time that they were being sizzled by the fire they were being mentally and physically tortured by large numbers of burning trees falling crisscross over the stream where they were lying, only the low banks preventing them from being crushed to death.

Although the trails were all closed by thousands of fallen trees, one of Bell's men who was least injured worked his way over hot ashes and fallen timber for 17 miles into Wallace and reported conditions. We immediately secured two doctors and a crew of men who shouldered packs of medical supplies, blankets and provisions and made their way to the burned men. A day and night crew were immediately set to work cutting out the trail leading from Wallace to Big creek, which was completed in four days, when the injured men were brought to the hospital on horses.

Ranger Hollingshead had a crew of 60 men on Big creek, about 22 miles southwest of

Wallace. On the afternoon of August 20, new fires were starting all around him and he gathered his men together and told them that it was time they were getting into a safer place and that it would be safer for them to travel through the recently burned over area south to the St. Joe river than it would be to go north into the green timber, on account of the possibility of fire spreading over the unburned region north of them. Forty-one of his men accepted his advice and followed him through the burned area and while there was still great danger from the fires and falling snags, with the exception of having badly burned feet and being nearly dead from exhaustion they arrived safely at the St. Joe river the next morning. The 19 men who refused to follow the ranger's instructions went north to what was locally known as the Dittman cabin, where there was a small cleared area around the cabin in the midst of heavy timber. Here they decided to make their stand as there was a wall of flames in every direction. They carried tubs and pails of water into the cabin with which they hoped to keep the cabin from burning. When the fire neared the cabin the heat became so intense the men were compelled to go into the cabin, which soon caught on fire and the small amount of water which they had stored was of little value. The men remained in the cabin until the burning roof fell in, when they rushed out hoping to find some means of escape, but 18 out of the 19 perished within 100 feet of the cabin, where their bodies were so severely burned that they could not be recognized. When the bodies were found a few days later the 18 dead men, five horses and two black bears were all close together.

One man happened to find a weaker place in the wall of flames than the other 18 and made a miraculous escape. He stated that he stumbled and fell right in the flames but somehow rolled through. After getting out of the fire he rolled on the ground to extinguish his burning clothing. His appearance after spending six weeks in the St. Joe hospital indicated that practically all of the skin had been burned off his hands and face. Without anything to eat he wandered through the burned area in a southerly direction for three days, coming out near St. Joe more nearly dead than alive.

(Opposite) GROVE FIRE, Prince George, B.C. — 40,000 acres burning. (B.C. For. Serv. photo).

Ranger Debitt had a crew of 100 men working on Setzer creek, north of the St. Joe river. He warned them of the great danger they were in and instructed them to go out to the St. Joe river. All of the men but 28 went out and were safe. The 28 men who remained were all burned into an unrecognizable mass and were later temporarily buried where they fell.

Ranger Rock had a crew of 125 men between Wallace and the St. Joe river. When they found they were surrounded by fire they located in what they thought was a suitable place and backfired and burned over a large area round about them. When the main fire approached, the fire again passed over the area that had been backfired, but the first fire had taken much of the fuel, therefore the second fire was not sufficiently severe to do great harm. The crew were all saved but one man, who shot himself presumably on acount of an insane fear of burning to death.

Rangers Danielson and Myers had a crew about 10 miles east of Wallace near the Idaho-Montana line, who also thought they would protect themselves from the flames by backfiring. But when the main fire approached it came with such force that it swept right over the recently burned area and soon reached the men. They had blankets and quilts with them which they threw over their heads and which more or less protected them from the flames for the instant, but the quilt soon caught fire and burned. One of the men had two woolen blankets which he separated and used to cover the whole crew, which consisted of 18 men. These woolen blankets served as sufficient protection to save their lives, with the exception of one who had evidently inhaled the flames and who fell dead on the spot where they had made their desperate stand. These men were so badly injured that they were carried out to the Northern Pacific railroad on stretchers. A special train awaited them and carried them to where a large trestle had burned out. They were carried around the trestle to another special train which brought them to the hospital at Wallace. Although they were all very badly burned their chief distress was in their lungs. They all finally recovered.

A large crew of men fighting that section of the fire near the Bullion mine on the Idaho-Montana line between Wallace and Saltese sought shelter from the roaring furnace in one of the tunnels of this abandoned mine. Those of the men who went in the tunnel sufficiently

HONKY TONK settlement of Grand Forks, Idaho, notorious camp when Taft Tunnel was being built. Entirely destroyed in 1910 fire, bars and bistros were back in business immediately after, tents set up in charred ruins (opposite bottom). (Ore. Hist. Soc. photo).

far to get beyond a ventilating shaft had little difficulty in surviving the fire, but the remainder of the crew, consisting of eight men who failed to pass the ventilating shaft were all suffocated.

The large crews of men under Deputy Supervisor Haines and Rangers Allen, Derrick, Kottkey, Fearn and Halm protected themselves through this terrible night by getting into the river or by backfiring.

Even though August 20 gave signs early in the morning it was going to be a very bad fire day, the fire which for several days had been held in check about five miles south of Wallace did not break over the lines to any extent until about 4 p. m. The O.-W. R. & N. railway officials, at the request of the citizens of Wallace and on account of information given by the Forest Service relative to the hazardous situation of Wallace because of the raging fires near-by, kept a train in readiness all afternoon so that at a signal to be given, of which the people were all familiar, the women and children could be removed from the city on short notice.

To become more familiar with the action of the fire on account of frequent requests from the citizens and railroad officials, I went up Canyon creek about 4 p. m. When I got about five miles up the canyon I noticed the fire had taken on new life and was burning at a terrific rate. Great columns of black pine smoke above the tops of the trees would explode into flame and send a swirling, swishing column of fire hundreds of feet into the air, A new fire was now raging on the hill close to Wallace, which indicated that the town was in great danger. I started down the canyon to notify the town that it was time for the women and children to go. About three miles out of town I met a man who was driving up the canyon to get his family, who lived on a little homestead about a mile farther up the stream. His progress was stopped by trees that had recently been blown across the road. He was ill and scarcely able to walk, therefore implored me to help him save his family. On account of the fallen timber I had left my horse on the Wallace side, so I ran back to the homestead, hoping to bring his family down to him. But when I reached the place several of my men had sought shelter there from the fire and, although the barn had already been burned, they were pouring water on the house. I knew from this and with the cleared space about the house that his family was safe, so I immediately

returned without them. Before I reached the man who was waiting for his family, however, the fire had swooped down the mountain with a roar that could be heard for miles. Great tongues of flames crossing the road cut me off from the man with the rig and my horse. Being well acquainted with the region, I knew of a small tunnel on a prospect about a half mile up the canyon from where the fire had crossed the road. I ran back as quickly as possible and entered the tunnel, the fire reaching the tunnel less than a minute after I entered. The heavy debris of brush and logs just outside of the tunnel set the tunnel timbers on fire and as they burned out the ground fell in, which, together with the strangling smoke, led me to believe I would be safer outside. There being a little water on the bottom of the tunnel I soaked my clothing throughly, held my wet hat over my face and rushed out between the burning timbers.

I quickly scratched a hole in a pile of sand just outside of the tunnel, in which I lay with my face down, which gave my face protection, but my neck and back and part of my head were badly burned. By midnight the fire had died down sufficiently to permit travel toward Wallace. I found the bridges all burned out and many places so hot that it was difficult to pass through. When I reached the city reservoir I found that a large pile of debris adjacent to the water main had set the outside of the wood stave pipe on fire. Believing that there was grave dan-

112

(Opposite and above) CHARRED EN-
TRANCE to War Eagle Tunnel on West
Fork, Placer Creek, where Ranger Ed
Pulaski and 40 men took refuge during fire.
Pulaski almost lost his life guarding tunnel
entrance in preventing half-crazed men from
breaking out to certain death or killing each
other to get at trickle of water on tunnel
floor. Pulaski suffered some loss of vision
in fire but lived to retirement, meeting
death in an automobile accident. A fire-
fighting tool invented by him is called a
"pulaski". (Ore. Hist. Soc. photo).

ger of the fire weakening the pipe sufficiently to cause a break, I carried water with my hat until the fire on the pipe was extinguished. My horse had torn loose but he had fallen in the fire a short distance from where I had left him and was dead. I reached Wallace at 2 a. m. and found the town on fire. Fully a third of the town had burned with a loss of more than a million dollars. When the hills adjacent to Wallace became a seething furnace and the town caught on fire, the women and children were loaded on the special train held in waiting, and taken out of town for the night, and brought back the next day. Another special train of box cars carried the patients from the Sisters' hospital. As they were taken out hurriedly many of them were taken without sufficient clothing to protect them, some being carried on stretchers. The two companies of federal troops who had been sent to Wallace to help fight fire rendered splendid service in helping to get the women and children on the train and in policing the town.

The results of August 20 and 21 were disheartening, but the fire was still burning in many places. New crews had to be immediately recruited, not only to fight fire but for rescue work. This was the hardest task of all. The men were afraid to go into the woods. On August 21 a fire was reported in the little north fork of the Coeur d'Alene river which contained a billion feet of white pine. A crew had to be sent there at once. By hard work it was secured and they had good luck in extinguishing the fire. Crews had to be organized to cut out the fallen timber from the trails, bury the dead, and bring out the injured. Runners had to be sent long distances to look up crews from which no report had been received.

One of these crews was led by Ranger Halm 75 miles up the St. Joe from Avery. The stations of Superior and Iron Mountain, Mont., on the Northern Pacific railway, were the nearest points to where this crew was working and these places were about 60 miles by trail. Three crews had been sent in from Superior and Iron Mountain Sunday, Monday and Tuesday, but they all returned without finding them, saying they could not get through the fire. We had learned that their entire pack train and one man had burned to death.

Deputy Supervisor Rosco Haines, who had originally taken the men to the head of the St. Joe river, was the only man who thought he knew how to get through the fires that were still

DEATH TRAP Roderick Ames and 5 members of Joe Bell's fire crew died in this hole on homestead of Joe Beauchamp on Middle Fork of Big Creek, Idaho, Aug. 20-21, 1910. A 35-year-old homesteader Ames had claim on Ames Creek, had come from Kellogg to bury his household goods to save them from advancing fire. Encountering Bell's large crew, he stayed to help. When inferno caught them, Bell ordered men to get in small pool in clearing on Beauchamp land. Seeing there was not room for all of them. Beachamp remembered cellar-like hole where he buried his valuables. 6 men huddled here and died of suffocation. (Ore. Hist. Soc. photo).

burning and find Halm's crew. So Haines, with two other men, started on Wednesday morning, August 24, for the head of the St. Joe. They traveled on horseback to the divide at the Cedar creek crossing, where they left their horses and started down the river on foot, frequently shooting as a signal to anyone needing help. The third day out, Friday afternoon, Haines shot a grouse, and Frank Mills, one of the crew, heard the shot, and the crew was found. On account of the thousands of fallen trees they were cutting their way out and had already come 20 miles. The crew had survived a terrible fire by getting on an island in the St. Joe river. Haines cut across country through fallen timber and burning snags, and after a long, strenuous trip reached a telephone that same night and phoned me at Wallace of the safety of Halm's crew — a more pleasing message has never been received!

The severity of the fire can be well known on account of the fact that in many places much of the green timber was actually consumed. This was especially true of the cedar. Also, a few days after the fire occurred, thousands of dead trout were found along Big creek and other streams where the fire was especially severe, this being caused, without doubt, chiefly by the large quantities of ashes getting into the streams, by the falling of burning logs, and the very heavy winds at the time of the fire.

By August 25 we had more than 100 men in the hospitals of Wallace and St. Joe. Many of them were in on account of smoke injury to their lungs, and only remained a few days, while others who were badly burned were in the hospital for as long as six weeks.

Up to 1910 the government had made no provision for the expenses of emergencies of this kind, therefore there was no way to pay with government funds the hospital bills which amounted to more than $5000. The Red Cross was good enough to furnish $1000 and the remainder was paid in full by donations from the forest offices throughout the service. The following winter Congress passed an act appropriating a small amount which was used in making small payments to those of the fire-fighters who were most severely injured and to the dependents of those who were burned to death.

The dead bodies were sewed up in heavy canvas and buried where they fell, as soon as possible, and later they were removed to St. Maries, Ida., and the plot of ground containing their graves permanently marked.

The difficulties to be overcome in paying off the men, many of whom were working under assumed names, identifying the dead men and finding out the names and location of their relatives and dependents, represented by practically all of the states and many foreign countries, culling out the false representations, and answering letters of inquiry from thousands of persons who had at some time lost track of an acquaintance or relative, produced a volume of work which took several years to clean up.

FIRE IN CARIBOO COUNTRY, B.C. after three days of rain. (B.C. For. Serv. photo).

HURRICANE AND FIRE made this havoc
in heavy stand of white pine on Little North
Fork, St. Joe River, Idaho, St. Joe National
Forest, in holocaust of 1910. (US For. Serv.
Missoula, photo).

FIRE NEAR DONALD. B.C. (B.C. For. Serv. photo).

WHITE PINE AND SPRUCE was burned
and swept to ground on Upper St. Regis
River in Cabinet National Forest — great
Idaho fire of 1910. (US For. Serv. Missoula,
photo).

FIRE NEAR GISCOMBE, B.C. (B.C. For. Serv. photo).

RANGER EDWARD PULASKI (left in group), descendant of Polish royalty, was credited with saving lives of 40 of his men trapped in Coeur d'Alene Forest by urging them to take refuge in mine tunnel on Placer Creek. Rear row, l. to r. — Pulaski, wife Emma, daughter Elsie, Gladys Noxon. Front row, Mr. and Mrs. Straw. (Ore. Hist. Soc. photo).

QUEEN CHARLOTTE ISLAND FIRE, on West Arm of Tuskatala Inlet, B.C.
(B.C. For. Serv. photo)

The Big Fire

Out of the underbrush dashed a man — grimy, breathless, hat in hand. At his heels came another. Then a whole crew, all casting fearful glances behind them.

"She's coming! The whole country's afire! Grab your stuff, ranger, and let's get outa here!" gasped the leader.

This scene, on the afternoon of August 20, 1910, stands out vividly in my memory. The place was a tiny timbered flat along a small creek in the headwaters of the St. Joe River, in Idaho. The little flat, cleared of undergrowth to accommodate our small camp, seemed dwarfed beneath the great pines and spruce. The little stream swirled and gurgled beneath the dense growth and windfall, and feebly lent moisture to the thirsting trees along its banks.

For weeks forest rangers with crews of men had been fighting in a vain endeavor to hold in check the numerous fires which threatened the very heart of the great white pine belt in the forests of Idaho and Montana. For days an ominous, stifling pall of smoke had hung over the valleys and mountains. Crews of men, silent and grim, worked along the encircling fire trenches. Bear, deer, elk and mountain lions stalked stary-eyed and restless through the camps, their fear of man overcome by a greater terror. Birds, bewildered, hopped about in the thickets, their song subdued, choked by the stifling smoke and oppressive heat. No rain had fallen since May. All vegetation stood crisp and brown, seared and withered by the long drought, as if by blight. The fragrance of summer flowers had given way to the tang of dead smoke. The withered ferns and grasses were covered by a hoar-frost of grey ashes. Men, red-eyed and sore of lung, panted for a breath

Ranger Joe Halm, former football player at Washington State College, joined the U.S. Forest Service in 1909. Working in the raw wilderness near the headwaters of St. Joe and Clearwater Rivers during the great 1910 fire he and crew were given up as lost. Yet he lived to write of his experiences when he retired, reprinted here from Montana, *Autumn, 1960. He died in Missoula in 1965.*

of untainted air. The sun rose and set beyond the pall of smoke. All nature seemed tense, unnatural and ominous.

It had taken days to slash a way through the miles of tangled wilderness to our fire, sixty-five miles from a railroad. On August 18, this fire was confined within trenches; all seemed well; a day or two more and all would have been considered safe. Difficulties in transportation developed which necessitated reducing our crew from eight-five to eighteen men.

I had just returned after guiding our remaining packers with their stock to one of our supply camps, when our demoralized crew dashed in. Incoherently, the men told of how the fire had sprung up everywhere about them as they worked. The resinous smoke had become darker, the air even more oppressive and quiet. As if by magic, sparks were fanned to flames which licked the trees into one great conflagration. They had dropped their tools and fled for their lives. A great wall of fire was coming out of the northwest. Even at that moment small, charred twigs came sifting out of the ever-darkening sky. The foreman, still carrying his ax, was the last to arrive. "Looks bad," he said. Together we tried to calm the men. The cook hurried the preparation of an early supper. A slight wind now stirred the treetops overhead; a faint, distant roar was wafted to my ears. The men heard it; a sound as of heavy wind, or a distant waterfall. Three men, believing safety lay in flight, refused to stay. "We're not going to stay here and be roasted alive. We're going."

Things looked bad. Drastic steps were necessary. Supper was forgotten. I slipped into my tent and strapped on my gun. As I stepped out a red glow was already lighting the sky. The men were pointing excitedly to the north.

"She's jumped a mile across the canyon," said the foreman, who had been talking quietly to the men. Stepping before them, I carelessly touched the holster of the gun and delivered an ultimatum with outward confidence, which I by no means felt.

"Not a man leaves this camp. We'll stay by this creek and live to tell about it. I'll see you through. Every man hold out some grub, a blanket and a tool. Chuck the rest in that tent, drop the poles and bury it."

The men did not hesitate. The supplies, bedding, and equipment were dumped into the tent, the poles jerked out, and sand shoveled over it. Some ran with armloads of canned goods to the small bar in the creek, an open space scarcely thirty feet across. Frying pans, pails, and one blanket for each man were moved there. Meanwhile the wind had risen to hurricane velocity. Fire was now all around us, banners of incandescent flames licked the sky. Showers of large, flaming brands were falling everywhere. The quiet of a few minutes before had become a horrible din. The hissing, roaring flames, the terrific crashing and rending of falling timber was deafening, terrifying. Men rushed back and forth trying to help. One young giant, crazed with fear, broke and ran. I dashed after him. He came back, wild-eyed, crying, hysterical. The fire had closed in; the heat became intolerable.

All our trust and hope was in the little stream and the friendly gravel bar. Some crept beneath wet blankets, but falling snags drove them out. There was yet air over the water. Armed with buckets, we splashed back and forth in the shallow stream throwing water as high as our strength would permit, drenching the burning trees. A great tree crashed across our bar; one man went down, but came up unhurt. A few yards below, a great log jam, an acre or more in extent, the deposit of a cloud burst in years gone by became a roaring furnace, a threatening hell. If the wind changed, a single blast from this inferno would wipe us out. Our drenched clothing steamed and smoked; still the men fought. Another giant tree crashed, cutting deep into the little bar, blinding and showering us with sparks and spray. But again the men nimbly sidestepped the hideous meteoric monster.

After what seemed hours, the screaming, hissing and snapping of millions of doomed trees, and the showers of sparks and burning brands grew less. The fire gradually subsided. Words were spoken. The drenched begrimed men became more hopeful. Some even sought tobacco in their water-soaked clothing. Another hour and we began to feel the chill of the night. The hideous, red glare of the inferno still lighted everything; trees still fell by the thousands. Wearily, the men began to drag the watersoaked blankets from the creek and dry

them; some scraped places beneath the falled trees where they might crawl with their weary, tortured bodies out of reach of the falling snags. The wind subsided. Through that long night beside a man-made fire, guards sat, a wet blanket around their chilled bodies.

Dawn broke almost clear of smoke, the first in weeks. Men began to crawl stiffly out from their burrows and look about. Such a scene! The green standing forest of yesterday was gone; in its place a charred and smoking mass of melancholy wreckage. The virgin trees, as far as the eye could see, were broken or down, devoid of a single sprig of green. Miles of trees, sturdy, forest giants — were laid prone. Only the smaller trees stood stripped and broken. The great log jam still burned. Save for the minor burns and injuries, all were safe. Inwardly, I gave thanks for being alive. A big fellow, a Swede, the one who had refused to stay, slapped me on the back and handed me my gun. I had not missed it.

"You lost her in the creek last night. You save me my life," he said, simply. His lip trembled as he walked away.

The cook had already salvaged a breakfast from the trampled cache in the creek. Frying ham and steaming coffee drove away the last trace of discomfort.

"What are your plans?" asked the foreman, after several cups of coffee.

"First, we'll dig out our tent, salvage the grub, and then look the fire over. We'll order more men and equipment and hit the fire again."

Little did I know as I spoke that our fire that morning was but a dot on the blackened map of Idaho and Montana. After breakfast we picked our way through the fire to our camp of yesterday. All was safe. We moved the remaining equipment to the little bar. Our first thought was for the safety of the two packers and the pack stock at our supply camp. The foreman and I set out through the fire over the route of the old trail, now so changed and unnatural. With ever-increasing apprehension we reached the first supply camp where I had left the packers. Only a charred, smoking mass of cans and equipment marked the spot.

What had become of the men? Not a sign of life could we find. They must have gone to the next supply camp. We hurried on, unmindful of the choking smoke and our burned shoes. We came upon our last supply camp; this, too,

RANGER JOE HALM, shown at right, after 1910 fire in which he almost died. Near head-waters of St. Joe River, he forced his men into Timber Creek, restraining 3 at gun point who tried to outrun flames. Deaths of Halm and crew were reported in Portland *Oregonian*, Aug. 26, retracted next day when men were found safe. Left in photo is Forest Service cameraman. (Ore. Hist. Soc. photo).

was a charred, smoldering mass. Still no signs of the men. A half mile beyond we suddenly came upon the remains of a pack saddle; then another; the girths had been cut. Soon we found the blackened remains of a horse. Feverishly we searched farther. Next we found a riding saddle. With sinking heart we hastened on. More horses and more saddles. The fire was growing hotter. We halted, unable to go farther. We must go back for help and return when the heat had subsided.

Smoke darkened the sky; the wind had again risen to a gale; trees were once more falling all about us. We took shelter in a small cave in a rock ledge where the fire had burned itself out. Here we sat, parched, almost blind with smoke and ashes. . . .

After what seemed like hours, we crept out of our cramped quarters and retraced our steps. The storm had subsided slightly. If the remains of the trail had been littered that morning, it was completely filled now. We came to a bend in the creek where the trail passed over a sharp

hogback. As we neared the top, we again came into the full fury of the wind. Unable to stand, pelted by gravel and brands and blinded by ashes, we crawled across the exposed rocky ledges. I had never before, nor have I since, faced such a gale. On the ridges and slopes every tree was now uprooted and down. We passed the grim remains of the horses and supply camps. In the darkness we worked our way back over and under the blackened, fallen trees. . . .

By firelight we ate and related our fears as to the fate of the packers. As we talked, one of the men, pointed to the eastern sky, cried, "Look, she's coming again!" The sky in the east had taken on a hideous, reddish glow which became lighter and lighter. To the nerve-racked men it looked like another great fire bearing down upon us. Silently the men watched the phenomenon which lasted perhaps ten minutes. Then the realization came that the sky was clearing of smoke. In another brief space of time the sun shone. . . .

From Ranger Haines I heard the story of our packers. Shortly after I had left them they had become alarmed. Hastily saddling the fourteen head of horses, they had left the supply camp for Iron Mountain, sixty miles away. Before a mile was covered they realized the fire was coming and that, encumbered with the slow-moving stock, escape would be impossible. They cut the girths and freed the horses, hoping they might follow. Taking a gentle little saddle mare between them, they fled for their lives, one ahead, the other holding the animal by the tail, switching her along. The fire was already roaring behind. On they ran, the panting animal pulling first one, and then the other. Hundreds of spark-set fires sprang up beside the trail; these grew into crown fires, becoming the forerunner of the great conflagration. By superhuman effort they reached the summit on the Idaho-Montana state line. Here the fire in the sparse timber lost ground. On sped the men down the other side until the fire was left behind. Ten miles farther, completely exhausted, they reached a small cabin, where they unsaddled their jaded, faithful little horse, threw themselves into a bunk and fell asleep.

Two hours later the whinnie of the horse awoke them. A glare lighted the cabin. They rushed out; the fire was again all around them! They rescued the little horse from the already burning barn and dashed down the gulch. It was a desperate race for life. Trees falling above shot down the steep slopes and cut off their trail. The now saddleless, frightened little beast, driven by the men, jumped over and crawled beneath these logs like a dog. Two miles of this brought them to some old placer workings and safety. . . .

They had crossed a mountain range and covered a distance of nearly forty miles in the little over six hours, including their stay at the cabin — almost a superhuman feat.

Returning to Wallace, I learned that the outside world had suffered far more than we. Eighty-nine men had given up their lives in the great holocaust. The hospitals were overflowing with sick and injured. Hundreds had become homeless refugees.

Assigned the task of photographing the scene of the many casualties, I had an opportunity to observe the extent of the appalling disaster and to reconstruct the scene of the last, hopeless stand taken by those heroic, unselfish men who gave their lives that others might live. . . .

Decades have passed through the hour-glass of time and nature has long since reclothed the naked landscape with grass, shrubs and trees, but the great sacrifice of human life is not, and can never be, replaced or forgotten.

RUINS IN WALLACE, Idaho, during Coeur d'Alene Fire of August, 1910. (U. of Idaho photo).

CUT-OVER PINE burning in Oregon's lower Cascades and close up view of blaze in night. (Ore. Hist. Soc. photos).

HOT HAVOC in Klamath Falls area of
Oregon. (Ore. Hist. Soc. photo).

(Opposite) WALLACE, IDAHO — virtually
wiped out Aug. 20, 1910, as fringe disaster
of great Coeur d'Alene fire. It broke out
near *Times* office on Bank Street, between
7th and 8th, firemen powerless to check
flames. (Uni. of Idaho photos).

128

RUINS OF WALLACE. Opposite top shows gutted railroad station and tracks where relief trains evacuated hundreds fleeing flames. Bottom left, presumed all left of hotel. Above, only unburnables left of Coeur d'Alene Hardware warehouse. Below was scene of flaming night — sparks, brands, cinders, smoke and falling timbers. (Univ. of Idaho photos).

Valley of Death

An uncertain sun, struggling through the pall of smoke that covered most of Western Washington and extended far out beyond the Columbia bar and cliffs of North Head, revealed in faint detail the remains of horror. The lush and beautiful valley of the Lewis River, flowing into the Columbia at Woodland, was now a blackened waste of charred trees, stumps and flattened farmhouses, dotted with the bodies of humans, horses and farm stock. This was the bitter aftermath of the so-called "Yacolt Burn."

The sultry and hazy 17th of September, 1902, climaxed a week of the worst forest fires which white men ever saw west of the Cascade Mountains. The flames on this day struck so swiftly thirty-five people were caught as they fled and burned to horrible deaths. A dozen towns, camps and settlements in the path of the holocaust were laid into embers. Some farmers were able to escape but hundreds left desolation behind. Sawmills and shingle mills every ten or fifteen miles in the heavy woods burned their lives away and blew cinders and sparks so high they fell on the city of Portland. From Bellingham to Eugene smoky skies were split by the eerie light of 700,000 acres of prime timber burning fiercely, timber as tall as the Washington Monument, growing out of a fern-bound rankness. This was "The Dark Day".

All over the Northwest terror and panic ran on frantic feet. The fire covered a wide area north of the Columbia River yet skirted the town that gave the series of fires the name — Yacolt. Buildings there were scorched by the heat over a mile away. Areas as far as Grays Harbor and Puget Sound were covered by a suffocating canopy of smoke and many are the tales and legends rising from the wind-blown ashes.

It has been reported many times that a woman at a religious camp meeting near Woodland ran screaming around the tents, shouting that Mt. Tacoma (Rainier) was erupting, exhorting everyone to pray and run for their lives — "Glory to God — the last days have come!" Another story, written at the time by a Kalama newspaper man, concerned the tragedy of the Ira Reid and C. A. McKeen families.

With the McKeen baby and five other children, they were planning a picnic for an Uncle George Smith, newly arrived from Kansas. In the big Reid wagon all set out along the north fork of the Lewis River for Trout Lake near the snow-crowned crest of Mt. St. Helens. Smoke in the air had been so prevalent all year they could have shrugged it off as "another fire somewhere". Whatever their thoughts they slogged on over the rutty trail roads, passing the Yale post office and continuing for two miles.

They got no farther. From off to the east and down a draw a wall of flame came roaring with all the fury of a gale, trees exploding like gunfire and hurling burning brands in all directions. Death and annihilation came quickly. A set of iron wheel rims and a few black bones were all the evidence that life once moved here. The men had unhitched the horses before they made a run for the creek, remains of the animals found a few hundred feet away.

Star route mailman, Walter Newhouse, was working in the yard of his home near Speelyai Creek and jumped in sudden fright when he saw the wall of fire bearing down the mountainside. He lost no time in hitching his two ponies to the mail rig and racing down the road. Yet he could not go fast enough. When a search party was able to get into the timber a week later it found the horses in positions of flight but it was days before the body of Newhouse was found. He lay against a log in a gully, clothes and skin burned away, a short length of buggy whip in his hand.

When a neighbor, Joe Polly, heard the roaring and crackling of the approaching fire, the hail of embers on his roof, he was just able to reach the comparative safety of a clearing. His sister-in-law, Mrs. John Polly, was not. She lived in Yale and with her husband away at work in a logging camp, she fought her way out of the burning house, baby in her arms, leading

(Opposite top) WALLACE from water tank 22 days after disastrous fire, hospitals still crowded with casualties. Below, ruins of Pacific Hotel, Hotel Street from 7th. (Uni. of Idaho photos).

HANEY, B.C. — fire in timber of Abernethy-Lougheed Logging Co., 1929. Below, remains of sawmill at Darrington, Wash. (Uni. of Wash. photo).

her young brother by the hand. The bodies were found in that relation.

Another settler fled to his cellar but flames swept through the house, burning boards crashing down into the excavation, causing heat enough to melt even metal. Near Kalama, D. L. Wallace and wife were burned up in a tent while camping.

On the Oregon side of the Columbia the Hamilton brothers burned to death in the sawmill settlement of Palmer and Jeff Eberhart at Bridal Veil saw the planing mill burning, drove a locomotive across a flaming trestle, returned to rescue another but was caught and badly burned, yet living to tell of his experience.

Back in the Yacolt area, near Ariel, Homer McGee also lived through a day of terror while his two brothers perished in the inferno. He was driving a team hauling firewood for the logging camp cookhouse when he caught the warning of the approaching fire. Dumping his load he started the team back up the rutty road. The smoke grew thicker and air hotter and suddenly at the bottom of a narrow ravine the full force of the racing fire came down with a deafening "whush". Even the stream seemed to be on fire. He tried to get the team turned around but had no time to do more than jerk the lines. He jumped headlong into the water already hot to the touch and buried himself as deeply as he could under a muddy overhang of roots. He prayed, living a horrible, searing lifetime in the minutes it took the worst of the blast to pass over him.

The aftermath was something he always tried to forget but never could. When he dared crawl out over the hot ashes and still glowing cover, he saw the charred horses and wagon iron. He groped his way along the edge of the river to the log landing, marked by the twisted piles of steel that had been the donkey engine. Some of the men had escaped, he learned. Eight others, including his two brothers, had been burned to shapeless heaps.

GOLD STREAM, B.C. fire still burning but
under control — 1958. (B.C. For. Serv.
photo).

CAMPBELL RIVER, B.C. — fire burning inside fire guard, looking toward Beavertail Lake, June, 1951. B.C.

HOT SPOTS in fire at Mechosin, B.C. after it had been confined — 1951. (B.C. For. Serv. photo).

ELWHA RIVER timber burning south of
Port Angeles, Wash. (Right) Ridge fire
burning through a tangle of windfall. (Uni.
of Wash. photos).

YOUNG SPRUCE burning at Frogmore Lake, B.C. (B.C. For. Serv. photo).

Tillamook Disaster

The 1902 Yacolt Burn and that in the Coeur d'Alenes in 1910 took the greatest toll of life in Northwest forest fires but more timber was destroyed in the great Tillamook Burn in 1933 — a half-billion feet of merchantable fir, hemlock and cedar.

Early on the hot, dry, morning of August 14, a sharp northeast wind swayed the trees in much of western Oregon including those in Gales Creek Canyon. Logging camps pulled in rigging crews, pulled off caulked boots and cut off steam — all but one. It disregarded dangerous conditions and kept on yarding logs. Late in the afternoon the donkey brought in one turn that slammed down across a tinder-dry cedar — and the rest is history.

A hundred experienced fire fighters went in to help the logging crews in Gales Creek Canyon and by evening had checked the fire not at all. Another fire was reported to the south and truck loads of men were sent to fight it, aided by CCC youths from Forest Grove.

For over a week soot-blackened men staggered away from the fire lines, replaced by others with only a few hours' sleep. The town of Forest Grove assumed all the bustle and excitement of a wartime outpost. The Army set up a supply system to get food and equipment to the fire crews, trucks rumbled in and out day and night. State and forest agencies recruited all available men, the USFS and Experiment Station directing all work.

On the 24th the fire "blew up", leaping through the tops of 400 year old trees on a 15-mile front, a pall of smoke hanging 9,000 feet up. Finally a fog blanket drifted in from the sea to bring the fire down to the ground where it was finally put under control.

All the sawmills, lath and shingle mills and pulp mills in the U.S. used approximately 12½ billion feet of timber in 1932. In ten days the Tillamook fire destroyed a like amount.

FIRE ON THE RAMPAGE in Oregon. (US For. Serv. photo).

Copyright, 1907, by
Barnard's Studio

Black Bear

BLACK BEAR, IDAHO, before fire of
May, 1908. Opposite, views of town after
wind swept fire up canyon. (Uni. of Idaho
photos).

OREGON FIRE, presumed near La Pine.
(Ore. Hist. Soc. photo). Opposite, British
Columbia blaze seen from Fregmore Lake
road. (B.C. For. Serv. photo).

146

FIRE NEAR DONALD, British Columbia.
(B.C. For. Serv. photo). Opposite top, last
man out of camp strides solemnly as fire
bears down on buildings. Bottom, remains
of railroad trestle after fire sweeps through
timber. (Ore. Hist. Soc. photos).

Death of a Forest

Beautiful Selway National Forest, said government officials in 1935, is no more. Ravished by one of the most devastating fires in two decades which laid waste to 240,000 acres, Selway as a forest became only a name, unburned portions added to adjoining forests.

It was a story of fire demons on the rampage, of the death of five men, of half a million dollars fire fighting cost and far more in loss of timber and young forest growth, of untold loss of fish and game. It was a story of gallant effort by the Forest Service, battling the blazes for 64 days and nights and when other means were exhausted, fighting fire with fire.

During the night of July 10, 1934, lightning struck in six of the forest areas. Men and equipment were rushed to the remote spots but a great wall of flame swept to the north and east, forming a 20-mile wide front. At the peak of the battle 5,000 men were strung along it and over 400 miles of trenches constructed, yet again and again the fire leaped over them and raced on. Every fire fighting device, every piece of equipment the Forest Service could lay its hands on, was employed. At last a back-firing policy was adopted, dangerous unless carefully done and under scientific conditions.

The man-made fire, caught up by the breeze, leaped into the branches, became a raging crown fire, roared on to meet the other. The two walls of flame came together. The swath created by the back-fire was too wide and for lack of fuel the main fire subsided. In a few hours the crisis was past. The great Selway fire which had resisted all attempts of control for two months, had killed a forest but increased fire fighting efficiency.

PORT NEVILLE, B.C. fire in June, 1925, that destroyed logging camp, driving men to boats. (B.C. For. Serv. photo).

STEAM AT BOTH ENDS as log train whistles for sawmill in British Columbia. (Crown-Zellerbach Canada photo).

150

The POWER
and the GLORY

With drums reverberating through the woods, replacing the stomps and snorts of oxen, and the blood-letting damnations of bullwhackers, logging became of age and began its power drive to high production. Steam and the demand for logs combined to get them off the ground into rivers, flumes and chutes. And sawmills dreamed of cutting a million feet a day.

The man who in the middle '80s first bolted a little single-cylinder vertical engine on the upper part of the slim, vertical boiler, connected it with a horizontal drive shaft and this with a vertical gypsy, started mechanical logging on its powerful way. When he bolted this crude contraption to a sled and hauled it into the timber, grizzled bullpunchers must have spouted disdain until they saw what this straining, jumping steam pot could do.

This first machine earned the name "vertical gypsy". The hauling line was small and pliable, 3/8 to ½ inch thick, the steam behind it less than 100 pounds. There was a small fairlead to guide the line onto the spool and the tender of it had to know his business or the line would pile up on the spool during hauling and either stall the engine on dead center or kink it into confusion. Either condition would prevent it from running free back to the woods when the line horse took hold of it.

The Dolbeer donkey represented a refinement of the vertical gypsy. It had the same vertical boiler but a horizontal engine, single cylinder, with a flywheel on the crank shaft, to overcome the tendency to stall and increase hauling power through slower gearing. The main shaft had a gypsy at each end. More donkey engine improvement came in the next few years. The gypsy was discarded for a wide-faced drum and double engines were developed. Machines were built with frames, sleds grew more substantial and steam pressure got up to 125 pounds. Rigging increased in size and hauling lines grew to a whopping big 1".

All this involved specially adapted valves, eccentrics, gears, friction blocks and pins, brakes and levers. When the 7x9 piledriver type of donkey came into use, with a spool on the end of the drum shaft for hauling wood logs or "parbuckle" timber on cars with the aid of the "gin-pole", the donkey "jammer" lorded it behind his levers and felt like a king. True, he had to have steam up by 6 a.m., fire his own pot all day and was lucky to have firewood sawed for him. His "pulls" were guided by a man or boy on a high stump gyrating his arms in imitation of a hooker back in the woods.

Then with the addition of a second drum to which the haulback or trip-line was attached, the faithful old line horse went down the skidroad to oblivion and logging entered its third stage of progress resulting in the several-speeds, many-drummed monsters of 60 tons and more with their devilishly ingenius efficiency. And always behind the feverish activity was the siren call — "More logs! Make 'er pay better! More logs — at less cost!"

LOG CHUTE of Hope Sawmills, Ltd. at Brookmere, B.C. (B.C. For. Serv. photo). Center, log train winds out of Washington woods, 1938. (Uni. of Wash. photo).

TWO-LOG TRAIL CHUTE showing details
of construction and corduroy road for trail
team in Big Creek, near Calder, Idaho,
1912. (St. Joe Nat. For. photo).

THE DONKEY PUNCHER *by* Charles O. Olsen

I'm a donkey-jammer, from hell and back
With a humboldt yarder or cracker-jack,
Or any old kettle that holds the fog
From leaking enough to haul a log.
That's me, you savvy, Old Pull-'em-quick,
I do the job, and I do it slick;
Whenever I open the throttle wide
You better look for a place to hide.

Send in your signal; I do the rest;
But shun the bight, if you know what's best.
I'll start her easy (account of the crew)
And watch for snags and sidewinders too.
I'll keep her bobbing, clear to the tree;
(A stump in the way is nothing to me);
I haul the log to the spar-tree top,
And let her flicker, to see her drop.

With me the choker is never foul,
And never a chaser lets out a howl.
I please them all and deliver the sticks,
And pile them up like a load of bricks.
If choker and line will stand the pull
I'll very soon have the landing full.
And like a flash, when I get the sign,
I go ahead on the "come-back line."

Of course, the hooker can boss the crew,
But I'm the guy that kicks her through;
When I am there she's always jake;
But how would it be if I wasn't awake?
Fireman, whistle-punk, chaser and all
Look to the donkey and hark to its call;
Noontime and night they hear me blow—
To eat their dinner, or homeward go.

Should things go wrong with the dear little pot.
I jump for the hay-wire, just like a shot;
I doctor her up with a liner, or grease,
Plug up her flues or her pet-cocks with ease;
I close the damper and watch her smoke—
To fire for me is simply a joke.
My wood-buck is loafing in the shade
To hear her snort as she takes the grade.

But a fellow can't make anything go,
With a hay-wire outfit and a dam' poor show—
Then it's toot! toot-toot! toot! toot-toot! toot!
You'd thing the punk was a'playing a flute;
You slam the throttle and cuss the crew,
And ball up everything that you do;
I'll tell the world, it's about as hard
As to look at a deuce, as your own hole-card.

DONKEY trailing logs, placed to service both lateral chute and hold back logs in main chute — in St. Joe National Forest after 1910 fire. (St. Joe Nat. For. photo).

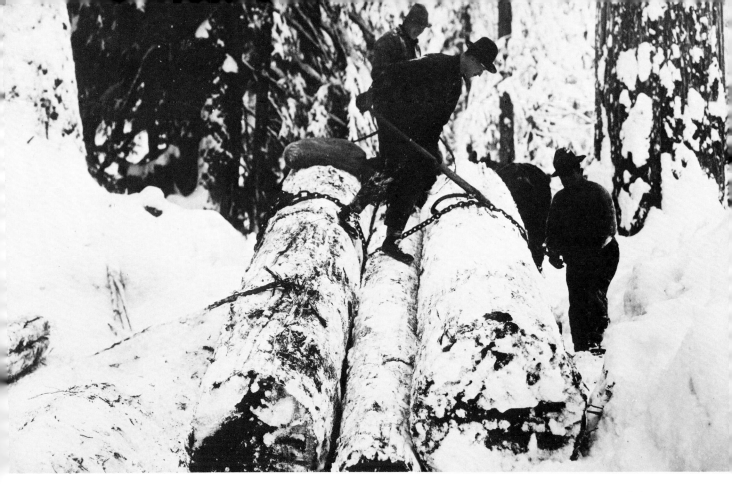

LOADING SLEIGH with short logs in British Columbia, 1912. (Vanc. City Arch. photo). Below, big log yarded by small donkey in Washington woods. (Uni. of Wash. photo).

LOG DRIVE on Oregon river. (Ore. Hist. Soc. photo).

LOG FLUME, Lumberton, B.C. (B.C. For. Serv. photo). Below, tractor replaces trail horses on this trail chute in Washington woods. (Uni. of Wash. photo).

FLUMING LOGS in British Columbia. Flume tenders stood on cat walk and patrolled sections of flume, checking for leaks and pile-ups. Center, B.C. Spruce Mills flume at Lumberton, B.C. (B.C. For. Serv. photos).

INSPECTOR RIDES FLUME at Yahk
Tie Reserve — C.P.R. operation, 1928. (B.C.
For. Serv. photo).

Skidroad Reverie

The '90s, that's a long piece back and the trail is covered with second-growth and cluttered up with piles of petty worries of push-button living. You can hardly find anybody who knew what First and Main, Cordova Street, Second and Burnside, Wishkah Street once were. Light the candles and sing a hymn — Old Man Blow 'Er In is dead!

Oh yes, there will be a few who remember the pioneering days of their youth when after long weeks of working their hearts out and sleeping in foul bunkhouses they came to town to go as far as their stakes would take them. They were drawn to the skidroad saloons, gambling houses, dance halls and "rooms upstairs" as surely as filings to a magnet. This was their home away from home. They crowded the massive, mirrored bars and drank with loggers, ranchers, railroaders and miners, with fishermen, sailors prospectors and prostitutes, with cowboys, adventurers from all over the world, with stakey men and stiffs. "Step up, boys. What you gonna have?"

If the logger was looking for a pal he would very likely see him slamming through some swinging door or standing at the mahogany with a big frothy scoop of beer in front of him. Anybody, anytime would sometime show up here where the whole world of roving labor passed on restless feet, stopping to watch barkeeps work, to join attentive groups around the gambling tables, talking in loud, assertive voices, singing the songs of a dozen tongues, arguing or lining the bars elbow to elbow, boasting of their exploits. They were the workers, doing the hard manual labor of the frontiers, on a temporary round of enjoyment and making the most of it while money and time was theirs. It was in such coin they paid themselves for months of enforced abstinence from social excitement, taking revenge for weary days of drudgery, drinking their fill of pleasure until the next time.

In the famous old saloons the drinking was boisterous and hearty for these were rough men on a spending spree. There was nothing calculated or furtive about it. A man blew in and if he saw no one he knew it was not five minutes before he had as many men calling him Shorty or Muley and setting up a drink for "this long-eared cuss from Christy's Landing." And once warmed he was buying drinks for the others. It was the spirit of skidroad lights and of a day when they were glittering.

The logger felt at home here. The bartenders or owners drifting in and out could tell a man with money at one quick blink and would let him "hoot and holler" as long as he could stand on his feet and pay for his drinks. They made him feel like one of the boys, that the gleaming brass rail and brass spittoons were put there just for him, that the reclining figure of the sad-eyed Venetian nude had been painted just for his yearning gaze. And he could look with fascination at the massed formation of odd-shaped, colored glass and pottery bottles of rye, gin, bourbon, rum from Demarrara and Jamaica, scotch from the land of kilts and bagpipes. Behind them, reflected in the spotless mirror, was the democratic show of highjinks and devil-may-care hooptedo, the glowing faces of his pals and their pals. On with the dance — let joy be unconfined!

Who can remember the ambrosial concoctions some of the mustachioed barkeeps whipped up out of those intriguing bottles? Pass up the standard Manhattan cocktail with its maraschino cherry impaled on a toothpick and have a Benson Banger that made a man hold on to his head to keep it from taking flight like a fool hen. And on those frosty mornings there was a warmer-upper that would have thawed out that mythical frozen logger. And there was the golden sheen of the jolly, frothy Tom and Jerry mixture that mellowed the mood, put blarney on the tongue, enraptured the senses and enthralled the spirit until the whole world was truly a place of good will to men.

Does memory bring back those free lunches, savory and salty to entice a man to eat and nicely calculated to make him drink? There were fish of many kinds, oysters and crab legs, meat balls, fish balls and "balls that were no balls at all," steamed clams and clam broth, stews, soups — it all brings tears to the eyes now.

And as the drinking was plunging and reckless so was the gambling. At night when the lights in the crystal chandeliers were bright and everybody was jumping or settling down under a full head of steam, then were the dice cavorting most gaily on seven and eleven legs and straights beat anything if a man could show them fast enough when eyes were bleary. There were plenty of sheep to be sheared — from the woods, grading camps, ranches and mines. They supplied the fleece that kept the spoilers in comfort. Fourth of July and Christmas holidays were their best times, with the wool heaviest and easy to sheer. The workingmen of the frontier were generally heedless, generous players and easy losers. For most of them this was a blessing in disguise. Far better to go broke in one glorious, meteoric orgy of a single night than to squander a stake on a continuous drunken blow out lasting three weeks. Better king for an hour than back to the old ten hours with wild glory booming in the ears, than never to have swigged at life's brimming cup.

There were rummy joints on the skidroad that offered a man not much more than a place to swill down whiskey, stomp his calks on the floor, with or without sawdust, squander a stake, meet a floosie and wake up alone in a crummy room, poke and pride gone with the night wind. There were cobwebby, smoky cellars where a man could get the "biggest scoop of beer in town for a nickel" and then have to fight his way out to daylight. Hard times and hard men!

There was no enchantment in those places. A man had to be a first class gent and find his fun in a first class house to get the color and glamor he came to town for. Where else could he hear a strident voice calling his name and realize this was not a gut-cussing shift boss talking to him but a man with a big-toothed smile and oiled hair putting out a paw to shake and asking, "How you been making out in the woods, friend? You're up at Conner's Inlet, aint you? Well, you're sure welcome here, big boy. What are you drinking? Step over here and have a drink on the Silver Seagull!"

Ah, wilderness were paradise enow.

LOGGING CREW at Burnaby, B.C. in 1900 (Vanc. City Ach. photo).

Daring Young Man in the Tree Tops

When the first logger hung lead blocks in a tree high above the ground, ran his hauling cable through them and out to the down logs in the woods, the era of high-lead logging was on its way. The donkey engine pulled the line, the noses of the logs were lifted high over obstructions, rear ends dragging easily. Logs were now in the air and ground lead methods were on their way out with oxen, horses and high wheels.

A tree rigged for such yarding was called a spar tree and it served especially well to transport logs over wide and deep canyons from one hill to another. A cable or tight-line was strung between spar trees standing on opposite hills, a traveling carriage operated by the hauling cable rigged on it, and the logs fastened to the carriage by steel slings or chokers, moved across, suspended in mid-air. The trees chosen as spars were big and sturdy, often 6 to 8 feet through at the butt and 250 in height. They must be trimmed of limbs, topped, guyed and rigged with hauling gear, this work done by skilled highclimbers or high-riggers.

A highclimber's equipment consisted of a pair of lineman's spurs and a very wide safety belt to which was attached a steel core manila rope ending in a slip knot. The steel core was a guard against the rope being cut by an accidental axe stroke, the slip knot for convenience in securing himself while working. His tools were an axe, cross-cut saw, flask of coal oil for cutting pitch and a small wedge or two to drive into the saw cut to prevent it from binding. Ax and saw dangled below the highclimber as he went up, attached to some part of his body by cords.

To climb he passed the rope around the butt of the tree and with free end in hand, stepped up the trunk, jerking the bight of the rope with him. He lopped off each limb as he met it, on a tree like this the lowest being perhaps 100 feet from the ground.

HIGH CLIMBER in British Columbia timber geared for trip up tall fir with safety axe, climbing spurs and rope. Tree was selected for height, tensile strength and sparsity of branches. (Crown Zellerbach Canada photo).

The height at which topping was done varied, usually about 200 feet where the tree trunk is 2 feet or more in diameter. The high-climber started work here by chopping a notch or under-cut in one side of the trunk, sawing a back cut on the opposite side. If the saw pinched, he drove in his wedges. When the top began to lean he withdrew the saw and let it drop to the end of the cord, digging his spurs into the tree and bracing himself for whatever was coming. For as the tons-heavy top, with its long, massive limbs dipped into space, a number of unexpected things might happen.

If he was in luck the top bent gracefully till it formed a right angle to the tree and plummeted to the ground. Then came the moments that tested the high climber's mettle, for as the top broke free, the kick-back set the tree to which he was clinging, in violent motion. It swung around fast and furiously in a circle that often reached 70 feet in diameter, as if some terrible power were trying to shake him loose. It was a sickening experience, highclimbers sometimes becoming nauseated, faint and hanging inert with death 200 feet below.

But topping was not always that tame. The trunk at the top might split and spread, drawing its enemy against the bark by the rope, squeezing him to death, or breaking his belt, or his back. To save himself he must let go hands and feet, trusting to the rope to hold him, let himself fall straight down the tree. If he is fortunate he escaped with bruises. Or he must throw off the rope belt entirely and cling with bare hands until the tree became steady again. Perhaps the top might not kick off at all, but slide backward just over the trunk to which he was fastened. In such a case he must circle swiftly to the side he judged the safest. If he made a mistake the falling limbs could brush him off. There were also accidents caused by carelessness, as when a highclimber unwittingly let go the end of his rope as he held it, or threw the bight of it over the top of the trunk after topping.

Descending was easier work. The highclimber jumped 6 or 8 feet down in a series of leaps, regulating the length of drop by flips of the rope. On the ground he laid aside his axe and saw and climbed again, taking with him a 15-pound steel pulley, a steel strap to hang it by and the end of a small steel cable. At the tree top he hung the pulley by the strap, passed the end of the cable through it and running it to the ground. By means of this gear all the rest of the heavy spar tree rigging was hoisted to him and he made it fast.

First the ends of half a dozen steel cables went up and were looped around the top to guy the tree to convenient stumps; more guy lines went around the middle of the spar to keep it from buckling when the high-powered, compound-geared donkey engine took a hard pull. Next the high-lead block was hoisted, weighing sometimes as much as a ton. Its sheave is often 3 feet in diameter, with an axis that ran on self-oiling bearings. The oil reservoirs, one on each side, may have held 15 gallons of oil. It was strung by a steel lead strap, much heavier than the guy lines. Sometimes the strap broke during a hard pull and the block hurtled down, burying itself in the ground. To prevent such accidents to the crew below, an extra safety lead strap was often used; with this the high-lead block was fastened to one of the guy lines. Should the main strap break the block would slide down the safety strap instead of plummeting straight to earth.

It usually took two to three days to rig a spar tree. Then the hauling cable, some inch and a half in diameter, was run from the donkey drum on the ground up through the high-lead block and out to the timber. A smaller cable was run from the drum along the ground and out to a lead block on some distant stump in the woods, from there doubling back to the spar tree. It was used to pull the main hauling block back to the timber after it had brought logs in.

Spar trees were rigged in all kinds of weather and to work high in the air in a rain storm was no picnic. Yet the kind of man that followed highclimbing did not mind. He was young, vigorous, thirsty for adventure, and was the most romantic figure in the woods. The tough veterans who did the falling, bucking, rigging slinging, choker setting or loading told many stories of the highclimber's exploits. They passed from mouth to mouth, from camp to camp, embellished by much picturesque language. Every

STARTING UP TREE high climber digs
spurs into bark, cinches up rope as bole of
tree grows smaller. (Crown Zellerbach
Canada photo).

section of the woods had its champion highclimber of whom it was very proud.

The work was more hazardous than most jobs entailing danger as accidents proved, but it carried the thrill for which the adventurous soul yearns. Hardship and danger were the spice of life to the highclimber and he took them in stride. There are many tales told of stunts performed in a spirit of bravado, to show off skill or win a bet.

Of one highclimber who later fell to his death, it was said that after topping a spar tree he always climbed to the top and stood on it smoking a cigarette, looking nonchalantly around the woods. The top of a tree trunk 24 inches across and 200 feet high is not a stable place to stand, especially if there is some wind blowing. Another daring highclimber stood on his head on the top of a spar tree to win a $25 bet. He performed the stunt in full climbing togs, spurred, belted and roped, with heavy logging boots on his feet.

The achievements of the highclimber seldom came to public notice, the logger working in the seclusion of the big woods, unseen, unapplauded. But among his fellows the highclimber held a position of great respect. His work was not constant but he must be always on the spot, ready to climb and replace or repair broken or faulty rigging. He was highly paid by the month whether he worked or not yet there was not too much competition for the job. He had to be a skilled woodsman and the years of his highclimbing competence were not many. The limitation of age and accident cut them short. He was not a good insurance risk — but who ever thought of that when he was young and going hell-for-leather?

NEARING TOP, which will be chopped off, climber has tightened rope through safety belt. Another 30 feet and he will secure himself to use both hands on axe. (Crown Zellerback Canada photo).

TIMBER! High climber braces himself for kickback of trunk which will buck and sway like a wild thing as 40 feet of top comes hurtling down. (Crown Zellerbach Canada photo).

A Tow to Port Moody

The story of towing logs in the Gulf of Georgia, Puget Sound and even deep-sea towing from The Queen Charlottes is that of battle with wind and tide, of towboat skippers risking rafts to supply mills needing the logs. They must smell a course through fogs, know every trick of the weather and the waters, know when to lay in a cove and when to buck a gale.

In the Pacific Northwest there are stretches of open coast exposed to the full fury of the ocean where tugs and rafts face peculiar hazards which practically prohibit the use of the ordinary flat boom; and even specially constructed rafts, secured with the strongest lashings of wire rope, sometimes come to pieces in heavy seas or break away from the tugs and are lost.

The simplest form of log boom used in Northwest waters is the bag boom, in which a number of logs without regard for arrangement, are encircled by a chain of boomsticks; but this is seldom used except as a temporary expedient in the most sheltered waters.

The flat boom is more generally used for towing in most inside waters. It is usually made up by men at logging camps using pike poles, pushing the logs inside two floating chains of boomsticks properly spaced between lines of piling. The logs are arranged in a close and orderly manner parallel to the sides of the boom. Other boomsticks are secured across the ends of each eight or ten sections. The usual width of a boom is sixty-six feet which is the standard length of boomsticks and swifters. The swifter sticks are laid across the top of the rafts at the end of each pair of boomsticks with short chains and toggles, connected in the same way as the boomsticks. A small donkey engine on a raft is usually used for hauling the swifter sticks across the boom. The space between swifters is called a section and though the number of feet of logs in a section varies considerably with the size of the logs, about 40,000 board feet is considered a fair average for fir logs. The average tugboat handles twenty to twenty-five sections on long tows and some of the more powereful tugs handle tows of forty to fifty sections, which would make a boom over half a mile long.

As the ordinary speed of a tug with a flat boom of logs is 1½ to 2 miles an hour in smooth water, and such booms cannot be depended upon to hold together in heavy weather, the tugboat captain must rely to a great extent upon his judgment of wind and water. Should the latter get up before he reaches a sure haven, a large part of the boom may be lost. Even the swell of a passing ship may cause some of the logs in a boom to ride on top of the boomsticks, and allow other logs to float out, while an unexpected tide rip may empty a boom of logs quicker than half a dozen men could in a sawmill boomyard.

Yet the flat boom is a convenient one for getting logs to the mills in sheltered waters where the towing distance is not more than twenty miles, losses insufficient to counterbalance the advantages. Even when a boom does break up it is often possible to round up most of the scattered logs or salvage them off the beaches when weather permits.

Running through narrow reaches of water, even in Puget Sound, has always been hazardous, notably Deception Pass where a 12 mile an hour tide sweeps along between Fidalgo and Whidbey Island. Northbound tows must run into Coronet Bay, at the upper end of Whidbey, and wait for the exact combination of wind and tide which will let them make the mile run through the pass. Tows have waited there as long as three weeks. They must catch a high tide when it turns to the ebb, there must be a little wind and a minimum swell. When the wind is down and tide is high the ebb flow will level the swells that run up from the Straits.

Many a tow however has piled up on these rocky shores. As many as ten tugs have been lined up with their rafts waiting for a favorable time to charge through the pass. Usually only four can make it with one ebb tide.

Southbound tows must catch the low water slack when the swells are down. The terrific

MAKING UP RAFT in British Columbia's Jervis Inlet. (Crown Zellerbach Canada photo).

SAWMILL BOUND log raft being towed
into Vancouver, B.C. Tug in foreground is
maneuvering to swing tail of raft into proper
position. (Crown Zellerbach Canada photo).

RAFT OF FIR LOGS being towed to Everett, Wash. sawmill for distribution in pond. (Uni. of Wash. photo).
Below, water lilies, yet? Logging roads vein British Columbia hillsides to boom. (Crown Zellerbach Canada photo)

CEDAR BOOM in Frederick Creek, B.C. (B.C. For. Serv. photo). (Opposite bottom) Towing spruce in narrow Siletz River, Ore. (Uni. of Wash. photo).

drive of a full flood tide carrying on an entire tow will submerge the head sections when they hit the slow slack water beyond the pass. Whole rafts have piled up on Strawberry Island with a thunderous crash of big butts splitting into slivers on the sharp rocks. The salvaging of such a wreck would take weeks, and the log pirates gathering up the strays.

Deep water presents many other towing problems. Most of the spruce cut on the Queen Charlotte Islands has historically been towed to Vancouver Island and mainland mills. Hecate Strait and Dixon Entrance are comparatively shallow, and as fierce gales sweep across them unchecked from the open sea, the forty miles or more of open water puts a severe strain on towboat skippers.

Again, all that part of the coast north of Queen Charlotte Sound has paid toll to the sea for logs towed across the thirty miles of open water at the Sound. A few flat booms have been towed across it during long weeks of calm weather but there is always risk from the ground swell alone.

Such hazards gave rise to the deep sea raft or crib, such as the Davis raft, a huge fabric of logs bound together with wire rope. A donkey engine is used to pile the logs together and haul the lines taut and when completed it is one gigantic bundle, often several hundred feet long and probably containing a million board feet, although some were built twice that size.

A base of logs with interwoven lines is formed, into this other logs piled with some overlapping for strength, then a final lashing of wire cables.

Many towboat captains have related thrilling experiences in riding out gales at sea with one of these great rafts, using it as a sort of sea anchor at times. Sometimes the rafts were lost and tugs forced to find safety. On a few occasions the rafts were picked up undamaged.

The chief difficulty with the big rafts is the slowness of progress through the water. Under calm conditions about a day is required for a powerful tug to take one of the rafts across Hecate Strait; and since on this run they do not take the risk of venturing out in stormy weather there have been times when much expense has been incurred by tugs standing by a raft for many weeks waiting for the chance to make the passage safely.

173

Building Ocean-going Log Raft on Hood's Canal, Washington.

OCEAN-GOING RAFT being built on Washington's Hood Canal. (Uni. of Wash. photo). (Opposite top) Building log rafts at Rock Bay, B.C. to be towed to Vancouver. Bottom, raft passing through Lake Washington Ship Canal locks, Seattle. (Uni. of Wash. photo).

Building Log Rafts at Rock Bay, B.C. to be towed to Vancouver

C24355B

(Above and opposite top) DAVIS SEA-GOING RAFT of spruce logs being built at Queen Charlotte Islands, B.C. in 1940. (B.C. Prov. Arch. photos). Bottom, Davis raft of 1¼ million feet of Sitka spruce in Selwyn Inlet, Queen Charlotte Islands, B.C. (B.C. For. Serv. photo).

BOOM MAN tackles tough job of getting center logs out as ordered by sawmill. (Uni. of Wash. photo).